The Future of American Agriculture as a Strategic Resource

The Conservation Foundation is a nonprofit research and communications organization dedicated to encouraging human conduct to sustain and enrich life on earth. Since its founding in 1948, it has attempted to provide intellectual leadership in the cause of wise management of the earth's resources.

The Future of American Agriculture as a Strategic Resource

Edited by
Sandra S. Batie and Robert G. Healy

The Conservation Foundation
Washington, D.C.

THE FUTURE OF AMERICAN AGRICULTURE AS A STRATEGIC RESOURCE

Cover and book design by Ronald L. Hibner

Typeset by United Graphics, Inc., Fairfax, Virginia

Printed by Edwards Brothers, Inc., Ann Arbor, Michigan

The Conservation Foundation
1717 Massachusetts Avenue, N.W.
Washington, D.C. 20036

Contents

FOREWORD xi

ACKNOWLEDGMENTS xv

INTRODUCTION
**American Agriculture as a Strategic
Resource: The Past and the Future**
Sandra S. Batie & Robert G. Healy 1

THE UNITED STATES IN THE WORLD
FOOD ECONOMY 2
Food Power 6
Foreign Exchange 8
Food Aid 9

AGRICULTURE IN THE DOMESTIC ECONOMY 11

THE FUTURE DEMANDS FOR U.S. AGRICULTURAL
PRODUCTION 13

AMERICAN AGRICULTURAL PRODUCTIVITY:
LESSONS FROM HISTORY 16
Forces Determining Past Agricultural Development 18

AMERICAN AGRICULTURE: THE FRAGILE SECTOR? 23
The Concerns 25
The Agricultural Resource Base 27
Research and Productivity 30
Relative Price Increases 30
Crop Vulnerability 32
Other Factors Influencing the Future of American
Agriculture 33

CONCLUSIONS 35
REFERENCES 38

CHAPTER 1
Competition for Land and the
Future of American Agriculture
Philip M. Raup 41

A NATIONAL OVERVIEW OF LAND-USE CATEGORIES . . 44

CROPLAND SHIFTS BY REGIONS 45

A FUNCTIONAL SURVEY OF
COMPETITION FOR LAND 48
    Land for Highways 49
    Land for Reservoirs 52
    Urbanization 53
    Land for Recreation 59
    Land for Energy 60

THE CHANGING BALANCE IN INTERREGIONAL
COMPETITION FOR LAND 65
    Irrigation 66
    Grain Exports 69

SOME FUTURE PROSPECTS 74

REFERENCES 75

CHAPTER 2
Soil Productivity and the
Future of American Agriculture
Frederick N. Swader 79

SOIL PRODUCTIVITY 80
    Chemical Properties 81
    The Physical Properties of Soils 82

CONSTRAINTS ON MAINTENANCE OF
SOIL PRODUCTIVITY 86
    Soil Erosion 86
    The Impact of Soil Erosion 89
    Soil Compaction 94
    Crop Residues, Energy Production,
        and Soil Productivity 96

FERTILIZERS AND SOIL PRODUCTIVITY 100
    Impact of Fertilizers 105
    Improving Fertilizer Efficiency 106
    Alternative Sources: Livestock and Municipal
        Organic Wastes 107

PROGRAMS, POLICIES, AND PROBLEMS 108

CONCLUSIONS 112

REFERENCES 113

CHAPTER 3
Agricultural Research and the
Future of American Agriculture
Vernon W. Ruttan 117

SOURCES OF GROWTH IN AGRICULTURAL
PRODUCTION 118
    The Frontier Model 118
    The Conservation Model 119
    The Urban-Industrial Impact Model 120
    The Diffusion Model 121
    The High-Payoff Input Model 121
    An Induced Innovation Model 122
    Induced Technical Innovation in the United
        States and Japan 124

THE CONTRIBUTION OF RESEARCH 127

CAN PRODUCTIVITY GROWTH BE SUSTAINED? 138

A PERSPECTIVE 145

APPENDIX 1: The Oklahoma State–USDA Research
             Productivity Studies 147

APPENDIX 2: A Retrospective View of Alternative
             Output, Input, and Productivity
             Projections, 1950-75 150

REFERENCES 152

CHAPTER 4
Irrigation and the
Future of American Agriculture
Kenneth D. Frederick 157

HISTORICAL TRENDS 159

WATER AS A CONSTRAINT TO IRRIGATION 164
    Augmenting Water Supplies 173
    Future Water Use for Irrigation:
        Some Conclusions 175

CHANGING ECONOMICS OF GROUNDWATER
IRRIGATION 176

ENVIRONMENTAL CONCERNS 181

THE FUTURE OF WESTERN IRRIGATION:
SOME CONCLUSIONS 184

FURTHER THOUGHTS: THE NATIONAL
IMPORTANCE OF IRRIGATION AND
RESEARCH NEEDS 186

REFERENCES 189

CHAPTER 5
Energy Dependence and the
Future of American Agriculture
Otto C. Doering, III 191

WHERE WE ARE TODAY 192

SOME REASONS FOR CURRENT ENERGY-USE
PATTERNS... 196
    The Interrelated Nature of Agricultural Production 201
    System Considerations 202
    The Nature of Agricultural Energy Demands.......... 204

OPTIONS FOR CHANGING ENERGY'S ROLE IN
AGRICULTURAL PRODUCTION 206
    The Effects of Changing Energy Prices................. 214
    Technology 215
    Institutions and Change 216

AGRICULTURE AS AN ENERGY PRODUCER 217
    Alcohol from Grain 218
    Alcohol from Cellulose Materials 219

EXPECTATIONS AND PERSPECTIVE 220

REFERENCES 222

CHAPTER 6
Crop Monoculture and the
Future of American Agriculture
Jack R. Harlan 225

HISTORICAL BACKGROUND 225

GENETIC DIVERSITY OF AMERICAN CROPS 234
    Corn 235
    Soybean 237
    Alfalfa 238
    Wheat 238
    Cotton 239
    Tobacco 240
    Sorghum 240

Potato 241
Rice 241
Oats... 242

SOME BASIC CONCERNS 243
Causes of Genetic Uniformity 243
Reduction of Risks 244
Future Problems 248

REFERENCES 249

CHAPTER 7
**Climate Change and the
Future of American Agriculture**
Robert H. Shaw 251

CLIMATE TRENDS, PAST AND PRESENT 253
Possible Misinterpretations of Climate Trends 256
Climate Cycles 258
Variability of Climate 260
Expected Future Changes 263

HUMAN IMPACT ON CLIMATE 264
Deforestation and Desertification 265
Atmospheric Particulates 266
Ozone Content 267
Carbon Dioxide Content 267
Acid Rain 268
Weather Modification 271

EFFECT OF CLIMATE CHANGE ON YIELD 271
Alternative Weather Scenarios........................ 272
Crop Yield Variability in North America 276
Climate Effects on Worldwide Yields 277
Grain Production in the Year 2000 282

CONCLUSION 284

REFERENCES 289

THE AUTHORS 292

Foreword

We enter the 1980s with a new awareness of the importance of agricultural resources. At the same time, we have scarcely begun to adapt our values and our behavior to this awareness. As the United States matured and urbanized, we increasingly took our food and fiber production for granted. To deal with the issues of the 1980s, however, we must begin by recognizing the economic and strategic importance of agriculture.

Domestically, agriculture provides direct employment for 3.9 million Americans and products valued at over $120 billion annually. Agriculture has been a major source of productivity growth in our economy, improving output per labor-hour at a rate that, since 1960, has been nearly three times the rate of productivity growth in nonfarm business.

Internationally, the United States is the world's leading exporter of wheat, coarse grains, and soybeans. One in every three acres cultivated in America is planted exclusively for export. In 1979, agricultural commodities generated $35 billion in foreign exchange, enough to offset more than three-fifths of the nation's bill that year for imported oil.

Agriculture is a strategic as well as an economic resource. National security depends on security of raw material supply, particularly in food and fuel. The United States has the good fortune to be self-sufficient in most food and fiber needs. Prudence dictates, however, that we watch carefully the factors responsible for the viability of American agriculture, lest, as with energy, problems catch us unprepared.

During the 1980s, comparable expansion of food and fiber production seems unlikely. Yields of many of the major food crops—wheat, sorghum, soybeans, potatoes—have reached plateaus in recent years. Observers point to multiple obstacles to further growth. The most productive lands are already being used. Productivity growth appears to be slowing. Rising energy costs are constraining the use of energy-intensive fertilizers. These same costs are constraining irrigation, and overpumping of groundwater is reducing the availability of water. Climate fluctuations and air pollution are reducing yields of highly monocultured crops. Limited support for agricultural research is inhibiting technological innovations.

We have paid a high environmental price for agricultural expansion. The farming of more acres has increased the soil erosion that makes agriculture one of the main contibutors to water pollution. Soil particles, many of them carrying pesticide and herbicide residuals, not only degrade water quality but cause siltation that interferes with navigation and reduces reservoir storage capacity. As John Timmons of Iowa State University states, "We are, in effect, exporting our soil and water quality in the form of food and feed grains." Fence-row-to-fence-row cultivation has also brought the draining of inland wetlands, the uprooting of shelter belts that formerly reduced wind erosion and provided wildlife habitat. The mining of aquifers has caused land subsidence in some areas and saltwater intrusion in others.

As the public becomes aware of the environmental and resource problems associated with our agriculture, reconciling agricultural production and conservation seems destined to become a major task of the 1980s. How do we improve environmental quality and manage resources without inhibiting agricultural production? Do our efforts to expand production today jeopardize the resources on which we will depend for production in the future?

Curiously, the environmental and resource issues associated with agriculture were largely ignored by conservationists during the 1970s. These issues, so critical in the minds of early conservationists such as Fairfield Osborn and Aldo Leopold, somehow failed to receive priority attention from their followers.

Partly, I suspect, this is because the conservation community assumed that some of the problems had been solved. Erosion, it seemed, had been solved back in dust bowl days. Now we know better. We realize once more that erosion control deserves an important place on the conservation agenda.

Perhaps the lack of attention to wise management of agricultural resources stemmed also from the type of solutions sought during the 1970s. The '70s were a time preeminently of single-focus adversary approaches to environmental problems, of tightening regulatory standards, of reliance on the courts to make the standards stick.

These approaches are not suitable to manage agricultural resources in the 1980s. The choices needed to reconcile multiple public objectives are too complex for single-focus measures, and regulations imposed from afar have limited promise as means to reshape the working methods of budget-conscious farmers, whose hostility to controls and whose continuing political power are widely known. Happily, the task of managing agricultural resources in the 1980s will be informed by the experience of the 1970s, bringing renewed recognition of the importance of multiple public objectives, of incentives as well as regulations, of active participation in the management process by everyone who will be affected by it.

The Conservation Foundation seeks to call attention to emerging issues and thereby to influence the agenda of the conservation community and of the country. In recent years, it has focused much of its effort on the juncture of economic and environmental policymaking, seeking more effective ways to reconcile conflicting interests. The Rural Resources element of the Foundation's Land Program responds to both these concerns, reflecting the emerging importance of rural resource issues and the need to address those issues in ways that transcend the adversary stances of the 1970s. As part of its Rural Resources effort, the Foundation commissioned the papers incorporated in this book. They are intended to contribute to the rigorous factual analysis and public understanding that are needed as the basis for sound agricultural policy in the 1980s.

William K. Reilly
President
The Conservation Foundation

Acknowledgments

The papers presented in this book were originally commissioned as partial fulfillment of a contract from the Federal Emergency Management Agency (FEMA). Their financial support is greatly appreciated. In addition, each of the papers has been reviewed and edited by many helpful individuals. The entire manuscript was edited by Beth Davis of The Conservation Foundation. Helpful review and assistance were given by other members of the Foundation staff, in particular J. Clarence Davies, Robert J. McCoy, John R. Noble, and Carol Zabin. We also want to thank Tony Brown, Evelyn Delillye, and Zeny Scott, who typed the manuscript for final publication.

S.S.B.

R.G.H.

INTRODUCTION

American Agriculture as a Strategic Resource: The Past and the Future

SANDRA S. BATIE AND ROBERT G. HEALY*

Many factors influence a nation's position in the modern world. The strength and technology of its military forces, the capacity of its industry, the size and capabilities of its work force, the cohesiveness of its social structure, the quality of its political leadership, and the ties that bind it to allies are among the most obvious. Events of the last decade have focused attention on yet another element of national strength and safety—the control of vital resources.

One of those resources is food. All nations must feed their citizens. U.S. experience during the twentieth century also has demonstrated the importance of having a surplus of foodstuffs available to export to allies during wartime and to aid other nations in postwar relief and reconstruction. Recently, the ability of the United States to export large quantities of food, as well as such nonfood agricultural products as tobacco and cotton, has been enormously important in cushioning our economy from the effects of increasing prices for foreign oil. In 1979, U.S. exports of agricultural commodities totaled $35 billion, up from only $7 billion in 1970. The foreign exchange generated by these exports offset more than 3/5 of the country's 1979 bill for imported oil.

World export dominance by the United States in certain key agricultural commodities increasingly has been seen as a possible

*The authors gratefully acknowledge comments received from Robert Anderson, Malcolm Bale, William Boehm, Thomas Dunlap, Richard Gilmore, Barbara Huddleston, Wayne Rasmussen, and A. Ellen Terpstra. Unless otherwise cited, numerical data presented in this paper are taken from *Statistical Abstract of the United States, Historical Statistics of the United States,* or annual U.S. Department of Agriculture statistical series.

1

source of national economic and diplomatic strength. For example, in the 1979/80 fiscal year, U.S. exports accounted for 55 percent of all wheat and coarse grain moving across national borders anywhere in the world. The sale or donation of foodstuffs to third world nations has long been an element of U.S. foreign policy. More recently, the U.S. government deliberately withheld food supplies from the Soviet Union in response to the Soviet invasion of Afghanistan.

Agriculture is also of strategic importance within the domestic economy. Farming directly employs 3.9 million persons; altogether there are about 6.5 million people living on farms. The assets of the farm sector total $820 billion, a figure equal to the market value of all the securities listed on the New York Stock Exchange. Annual sales of farm products are over $120 billion. These products are eventually converted into $400 billion in consumer expenditures for food and clothing. For several decades, farming also has been a major source of productivity growth in the U.S. economy. Since 1960, output per hour worked in agriculture has risen by 172 percent, a rate of productivity growth nearly three times that in nonfarm business.

It is easy to underestimate or overestimate the importance of agriculture to the strength and prosperity of the United States. On one hand, the fact that the United States has never in its long history faced serious nationwide food shortages has led many people not directly connected with agriculture to take the national food supply for granted. On the other hand, the overwhelming dominance of the United States in world grain trade has led to widespread public belief that the United States possesses a potential "food weapon" comparable in power to OPEC's "oil weapon." Neither of these assumptions is correct. In reality, U.S. agriculture is an important source of national influence, but one that has quite limited diplomatic and political uses. U.S. agriculture is a marvel of the application of science and technology to production, but it is vulnerable to a number of identifiable threats.

THE UNITED STATES IN THE WORLD FOOD ECONOMY

The role of the United States in the world food economy reflects two facts: widespread self-sufficiency in food production by the world's nations, and U.S. dominance in food exports.

First, the great majority of nations can produce most of their domestic food requirements, particularly staple protein sources directly consumed by humans. Only a little over 1/10 of world food output is traded internationally. The Food and Agriculture Organization of the United Nations (FAO) estimates that in both 1963 and 1975, 76 percent of the world's population "lived in countries which were 95 percent or more self-sufficient in food energy" (FAO, 1979). In less developed countries, the bulk of the population lives on farms. Much of the food consumed in these countries never enters domestic trade channels, being eaten by the farm family or bartered within the village where it is produced. Most of the larger developed countries have their own highly sophisticated agriculture. Even highly industrialized countries, such as Japan and much of western Europe, have had problems dealing with chronic surpluses of certain agricultural commodities.

But most nations are not completely self-sufficient, and some depend quite heavily on food imports to supplement domestic production. A relatively small number of the food importers are very poor countries receiving food donations or concessionary sales; Bangladesh and Egypt are good examples. Even this dependence should not be overemphasized, however. O'Hagan (1976) notes that India, which imported massive quantities of U.S. food during the 1960s, was still 97 percent self-sufficient in kilocalories from 1961-1973. Many of the impoverished, food-short nations have good long-term agricultural potential, but they lack the infrastructure and political support for agricultural modernization.

By far the bulk of world food imports are consumed in countries that can afford to pay market prices for food. These include industrialized countries and some of the faster growing developing nations. Among the major nations most dependent on foreign food are Japan, the United Kingdom, West Germany, South Korea, Iran, and Saudi Arabia. Such countries import food for several reasons. First, they import crops that are not easy to raise locally or that serve newly acquired tastes in food consumption. For example, higher income groups in many Asian and African countries are adopting a Western style diet by greatly increasing consumption of white bread and meat. Therefore, these countries have begun to import large quantities of wheat flour, corn, and feed concentrates. Second, many countries have found that they can obtain some farm products at a lower cost from foreign suppliers than by producing them domestically. Thus, Japan imports soybeans; Germany imports wheat and animal feeds; and Iran buys wheat, rice, and meat. Finally, countries import to cushion their domestic economies from the effects of poor domestic harvests. Traditionally, many countries adjusted to poor grain crops by slaughtering part of their animal herds, thereby reducing the demand for feedgrains. But many are now reluctant to subject their consumers to the resulting wide swings in meat supplies. Thus, the Soviet Union adjusted to a poor harvest in 1972 with unprecedented purchases in the world grain market.

Given enough time, most of the world's food importing countries could become self-sufficient in food. But it would be expensive and inconvenient and would in many cases mean that their consumers would pay higher prices for food and would have a less varied diet.

During the last two decades, there has been some tendency for developed countries, as a group, to become more self-sufficient in food, while the less developed countries, again as a group, become slightly less self-sufficient. In particular, less developed countries have greatly increased their imports of grains, which rose nearly threefold between 1960 and 1977. The bulk of these increased grain imports have come from the United States.

This brings us to the second fact essential to understanding the world position of U.S. agriculture: U.S. export dominance. The United States is blessed, among all the nations of the world, with an exceptionally large and fertile land base, particularly when considered relative to the U.S. population. This land is especially suited to the massive, low-cost production of wheat, corn, and soybeans. According to agricultural economist Don Paarlberg:

> There is no area of the world so well endowed agriculturally as the United States. The American Midwest, stretching from the Appalachians to the Rockies and from Mexico to Canada, is unmatched in the world for its size, its favorable topography, fertility, climate and transportation, its agricultural institutions and the managerial capability of its farmers. The American agricultural system is naturally suited to competition in world markets [Paarlberg, 1980].

Table 1 shows how the United States, which contains less than 5 percent of the world's population, has used its agricultural bounty to produce 13 percent of the world's wheat, 29 percent of the world's coarse grains (mainly corn), 17 percent of the world's meat, and 62 percent of the world's soybeans. Most of the meat

Table 1.   U.S. Contribution to World Food Supply

(million metric tons)

	Production			Exports		
	World	U.S.	U.S./ World (percent)	World	U.S.	U.S./ World (percent)
Wheat and Wheat Flour 1976/77-1978/79	415.1	54.2	13.1	69.1	30.0	43.4
Coarse Grains[a] 1976/77-1978/79	718.5	205.3	28.6	85.2	53.3	62.6
Meat 1976-78	103.4	17.8	17.2			
Rice 1976/77-1978/79	368.8	5.2	1.4	10.7	2.3	21.5
Soybeans 1976/77-1978/79	72.0	44.6	61.9	22.1	18.3	82.8
Cotton 1976/77-1978/79	13.2	2.6	19.7	4.1	1.2	29.3

a. Corn, barley, oats, sorghum, and rye, excluding products.

produced is consumed domestically. For the other crops, the high rate of production relative to the rate of domestic consumption leaves huge quantities of crops available for export. Overall, it has been estimated that one out of every three acres of U.S. cropland produces for export. The United States dominates world trade in coarse grains and soybeans and is the most important single country in the world wheat market. Even for rice, a basic world foodstuff of which the United States is not a large producer, an excess of production over domestic consumption means that the United States accounts for 21.5 percent of world rice trade.

Besides the United States, there are only a handful of countries that can produce large surpluses of these commodities for export: Canada, Australia, France, and Argentina for wheat; France, Argentina, and Canada for coarse grains; Brazil for soybeans; and China and Thailand for rice. Some of these nations, moreover, have highly variable climates and rainfall and are not as consistent producers as the United States.

The United States also dominates world stocks of these vital agricultural products (see Table 2). Some of the stocks, owned by the government or by the large commodity trading companies, are held in warehouses. Much of the stocks are stored on the U.S. farms where they were produced. Despite much global alarm since the 1972-1974 world production shortfall severely pinched world grain stocks, no agreement has been reached on internationally controlled or internationally coordinated grain reserves.

These two facts—widespread national self-sufficiency and U.S. export dominance—define the role of the United States in the world food economy. They also define the strengths and limitations of U.S. "food power."

Food Power

The issue of U.S. food power was discussed widely during the early 1970s, when tight world grain supplies and the successful formation of the OPEC cartel caused some observers to believe that the United States possessed a "food weapon" comparable to OPEC's "oil weapon" (Butz, 1974). But even a superficial examination of the data shows that because most nations produce most of their own food "there is not general American 'food power.' Rather, there is American grain power" (Wallensteen, 1978). And even grain power affects only a handful of very poor countries somewhat dependent on U.S. aid and a larger number of wealthier nations that buy U.S. grain to upgrade already adequate diets or to support increased levels of meat production.

The United States does have considerable power—should it decide to exercise it—over grain supplies. This power is greatest

Table 2. U.S. and World Food Stocks

Spring 1980 (million metric tons)

	U.S.	World[a]	U.S./World (percent)
Wheat and Wheat Flour[a]	26.5	87.7	30.2
Coarse Grains[a]	53.5	89.9	59.5
Rice[b]	1.1	23.1	4.8
Soybeans	10.9	12.9	84.5

a. May not include entire USSR stock.
b. Stock data not available for all countries; Burma and China excluded.

in the short run. By withholding grain from world trade, the United States could make citizens of a number of countries modify their diet for a time. World grain prices would escalate and the wealthier grain-importing countries would bid remaining world supplies away from less affluent nations. But before too long, more countries would become self-sufficient, demand for grain would fall as other food products were substituted, and grain exporters outside the United States would take up some of the slack.

Even though most observers will agree that U.S. grain is not nearly so vital to the world economy as OPEC oil, there is considerable disagreement about grain trade's potency as a tool of diplomacy. Schneider (1976); Laird (1978); and Levine, Rushing, and Movit (1979) have pointed to grain sales as a possible source of diplomatic leverage, particularly over the Soviet Union, where U.S. grain has been used in a major effort to build up livestock herds and hence reach targets for domestic meat consumption. Morgan (1980) points out that American grain exports and OPEC's oil exports are both "the crucial increment that bridges the [world] gap between adequate supply and scarcity." He argues that the United States could exercise power in the grain market and raise prices by creating a national grain exporting agency and/or by organizing the handful of other major exporters in a grain cartel.

Other experts believe that U.S. grain power confers little or no diplomatic advantage. Robert Paarlberg (1978) says flatly that, "even under the unusual 1972-75 conditions of market scarcity, food was never a particularly useful instrument of diplomacy." Gilmore (1980) notes that the U.S. government has exercised little effective control over foreign activity by the large trading companies that handle over 80 percent of U.S. grain and oilseed exports. He points out that when U.S. longshoremen refused to load grain for Iran following the 1979 embassy takeover, "subsidiaries of American grain companies allegedly shipped Australian wheat and Canadian barley to Iran." A report prepared for the Congressional Committee on International Relations asserts that while the United States may have "opportunities to exercise food power over specific countries based on market control," any general use of U.S. food power as a diplomatic weapon would run counter to U.S. trade policy emphasizing free trade in agricultural products and may "engender trade retaliation in other areas" (U.S. Congress, 1977).

Foreign Exchange

The explosive growth that has taken place in U.S. exports of farm products since 1972 has been a major factor in helping this country pay for imported oil and other foreign goods. Table 3 summarizes developments in U.S. foreign trade over the last 15 years. Agricultural exports were steady until 1972-1973, when poor harvest in several parts of the world dramatically increased demand for U.S. grains, including a purchase of unprecedented size by the Soviet Union. Since that time, world demand for U.S. farm products has remained high, even in years of good foreign harvests. Despite the reduction in sales to the Soviet Union and to Iran, the U.S. Department of Agriculture (USDA) predicts that farm exports will still reach record levels in fiscal 1979/80.

One of the factors making the agricultural sector so important to the U.S. balance of payments is the fact that the United States exports many products (such as animal feeds) that are responsive not only to foreign population increases but also to income

Table 3. U.S. Foreign Trade, 1965-79

(millions of dollars)

	Agricultural Exports	Agricultural Imports	Petroleum Products Imports	Trade Balance (Agricultural)	Trade Balance (Nonagricultural)
1965	6,229	4,087	2,063	+ 2,142	+ 3,710
1966	6,881	4,491	2,127	+ 2,390	+ 2,134
1967	6,380	4,452	2,086	+ 1,928	+ 2,481
1968	6,303	5,024	2,345	+ 1,279	− 146
1969	6,022	4,957	2,560	+ 1,065	+ 534
1970	7,259	5,770	2,770	+ 1,489	+ 1,345
1971	7,693	5,823	3,323	+ 1,870	− 3,894
1972	9,401	6,467	4,300	+ 2,934	− 9,340
1973	17,680	8,419	7,614	+ 9,261	− 8,039
1974	21,999	10,247	24,210	+11,752	−14,748
1975	21,884	9,310	24,814	+12,574	− 2,833
1976	22,997	10,992	31,794	+12,005	−20,672
1977	23,636	13,439	41,526	+10,197	−39,972
1978	29,407	14,804	39,104	+14,603	−46,824
1979	34,745	16,722	56,046	+18,023	−46,741

growth. In contrast, the agricultural products imported by the United States are heavily weighted by commodities (such as coffee and sugar) for which demand is not very sensitive to income growth. Thus, U.S. agricultural exports have grown much faster than agricultural imports, creating a trade surplus that has partially offset the growing deficit in the nonagricultural trade accounts.

The largest individual customers for U.S. farm products are the European Economic Community ($7.7 billion in 1979); Japan ($5.2 billion); the Soviet Union ($2.8 billion); Canada ($1.6 billion); and Korea ($1.4 billion). By groups, 56 percent of U.S. farm exports go to developed nations, 33 percent to developing nations, and 11 percent to nations with centrally planned economies. Members of OPEC have shown increasing demand for U.S. farm products, but as of 1978 they still accounted for only $2.2 billion in purchases. Several of the most important petroleum exporters, such as Saudi Arabia, Kuwait, Libya, and the United Arab Emirates have small populations and will never be able to absorb large quantities of U.S. food. Thus, U.S. farm exports do not pay directly for oil imports but rather are sold to other oil importing countries. Consequently, the typical U.S. agricultural trading partner must generate foreign exchange for both oil and food imports. Nations such as Japan and West Germany have proved very adept at doing so, but many of the less developed countries have run large trade balance deficits, financing them with international bank loans. Because of this interdependency, the future course of demand for U.S. exports will be influenced heavily by the overall health of the world trading system, and particularly by the ability of developing countries to sell their own products on world markets.

Food Aid

The existence of U.S. agricultural surpluses has made possible food aid, either by gift or by concessionary sale, to poor countries. Since 1954, the United States has donated more than $10 billion in agricultural commodities and sold $17 billion for local currencies or on a loan basis. Food aid has had multiple—and sometimes conflicting—purposes. It has been intended to serve humanitarian ends, support U.S. diplomacy, cultivate future export markets, and dispose of U.S. crop surpluses.

The history of U.S. food assistance reveals the varied objectives that food assistance has served. The United States gave large quantities of food to India during India's famine years in the late 1960s and to Bangladesh in the early 1970s. Food was shipped in great quantities to Europe during and after World War II, and to Vietnam and Cambodia during the Vietnam war years. Today, for a combination of humanitarian and diplomatic reasons, Egypt is the largest single recipient of U.S. food assistance.

To a limited extent, food aid is a source of national influence. Cartons of commodities are labeled "A Gift of the People of the United States of America" in an attempt to create good will for this country among the impoverished recipients. It is unclear whether food aid has such an effect, in part because the very need to accept food aid frequently wounds national pride. Moreover, it is probably foolish to think that food aid can by itself change another nation's foreign policy. For example, in 1966-1967, when an estimated 60 million Indians were being sustained by U.S. food aid, the Indian government was a severe critic of U.S. policy in Vietnam.

Food aid also imposes moral obligations on the donor. It is easy to give food to the needy when huge surpluses are filling domestic warehouses. But in years of short harvests, food aid may come at the cost of higher prices for the U.S. consumer. A United Nations' report notes that "during the world food crisis of 1972-74, when the need was greatest, the volume of [world] food aid fell drastically from [almost 13 million tons annually] to under 6 million tons in 1973/74" (FAO, 1979, p. 197). Since then, international negotiations have taken place in an attempt to ensure a more steady flow of aid. But pledges to date are still short of the minimum goal set by the 1974 World Food Conference.

Food aid also has been criticized for offering recipient countries an excuse for neglecting development of their own agricultural sector (Seevers, 1968; Isenman and Singer, 1977). The availability of low-cost foreign food may depress retail prices below a level profitable for local farmers. Perhaps more important, it can give governments an excuse to neglect investment in agriculture in favor of more visible industrial projects, or to procrastinate on needed land reforms. Senator George McGovern notes a lack of coordination between food and development programs, observing that "what AID [Agency for International Development] seeks to do to increase food production in developing countries is sometimes undercut by USDA's shipments of large amounts of food aid to such nations and by U.S. trade policies managed from elsewhere in the bureaucracy" (McGovern, 1979). On the other hand,

D. Paarlberg (1980) points to six developing countries (Republic of China, Brazil, Iran, Peru, Chile, and Colombia) and two developed ones (Japan and Spain) as "examples of former [food assistance] recipients that became good United States commercial customers."

AGRICULTURE IN THE DOMESTIC ECONOMY

Despite export growth, agriculture's size within the domestic economy has diminished over time. In 1978, agriculture contributed 2.8 percent of the U.S. gross national product (GNP), down from 5.6 percent in 1950 and about 10 percent in 1929.* The proportion of the labor force engaged in agriculture has fallen even more drastically, from 21 percent in 1929 and 11.5 percent in 1950, to 3.2 percent in 1979. This decline in numbers, incidentally, along with post-1962 redistricting of state legislatures, has been an important factor in reducing the once potent political influence of farm interests.

Agriculture's importance to the domestic economy, however, is understated by GNP or labor force comparisons, for many of the products produced on farms are subject to further processing by other domestic industries. For example, the farm share of the retail price of a 37 cent loaf of bread is only 5.4 cents. The rest of the cost represents value added in the course of milling, baking, packaging, transportation, and wholesale and retail trade.

Agriculture is also an important market for products made elsewhere in the economy. Greater use of farm machinery, fertilizers, fuels, and insecticides has meant that an increased share of the total cost of farm production is spent off the farm. For example, farmers spend six cents of every dollar of their sales on chemicals and thereby consume about 12 percent of all the yearly production of the chemical industry.

Using input-output techniques, it can be calculated that each million dollars of grain produced for final demand will create jobs not only for 30 grain farmers, but also 32 other jobs, including 7 jobs in wholesale and retail trade, 2 in transportation, 4 in miscellaneous business services, 2 in chemicals, 4 in other manufacturing, etc. Jobs created in the "downstream" industries in which farm products are used can be calculated, albeit with some diffi-

---

*These data are based on "agriculture, forestry and fisheries," with farming making up the overwhelming bulk of the total.

culty. For example, for every farmer producing grain for bakery products, there are 64 jobs in other industries involved in producing those products, including 31 in the baking industry itself (U.S. Bureau of Labor Statistics, 1980). Each farmer also produces food for 65 U.S. residents, enabling them to use their own labor to produce goods and services elsewhere in the economy.

Over the years, among the most significant effects of agriculture on the U.S. economy have been those on productivity growth. For several decades there had been a substantial shift of labor out of farm employment and toward the industrial and service sectors, which historically had much higher average productivity. One estimate is that "during the first 2 postwar decades [this shift] contributed an average of 0.5 percent to the productivity growth rate of the economy as a whole" (National Center for Productivity, 1978). Between 1966 and 1976, however, the gap between agricultural and nonagricultural productivity had so narrowed and the number of agricultural workers had so diminished that the shift out of agriculture added only 0.1 percent per year to the productivity growth rate of the U.S. economy. In the future, the shift is expected to have only a negligible impact on total growth (Dennison, 1979).

Looking at agriculture only as a source of labor for the rest of the economy obscures the precondition that made that shift possible—the long record of sustained growth in productivity within the agricultural sector. In terms of labor productivity, output per hour worked in agriculture has risen at a compound rate of 5.7 percent per year, against 2.0 percent in nonagricultural private business. Since 1970, labor productivity elsewhere in the economy has increased at only 1.5 percent yearly while agricultural sector productivity has grown by a respectable 5.5 percent.

Part of the increase in labor productivity in agriculture has been achieved by substituting other inputs—machinery, fertilizers, herbicides—for human labor. But agriculture also has done well when its performance is measured in terms of "total factor productivity," that is, output per unit of *all* inputs. Between 1970 and 1978, this measure has risen at a compound rate of 2.1 percent yearly.

U.S. consumers have been among the biggest beneficiaries of this country's agricultural productivity. In 1977, expenditures on food accounted for only 18 percent of total consumer expenditures, down from 24 percent in 1950. Since much of the cost of food to the consumer includes costs of processing, transportation and marketing, and preparation of restaurant meals, the share of total consumer expenditures actually paid to the farm sector is only about 6 percent. Even while the proportion of income spent on food was falling, American consumers were enjoying an exceptionally wide range of dietary choice, paying relatively low prices for such former luxuries as out-of-season fruits and winter vegetables.

Residents of most other countries spend a significantly higher proportion of their consumer expenditures on food. Figures are not strictly comparable because of differing national statistical definitions, but the United Nations reports, for example, that residents of Italy devote 42 percent of consumer expenditures to food; Japan, 36 percent; Brazil (Sao Paulo State), 56 percent; India (urban), 61 percent; France, 34 percent (of income); and Switzerland, 22 percent (of income) (FAO, 1977). The ability of U.S. farmers to produce low-cost food is thus a major factor in maintaining the standard of living for all U.S. residents.

THE FUTURE DEMANDS FOR U.S. AGRICULTURAL PRODUCTION

Beginning early in the 1970s, U.S. agriculture appears to have entered a new era characterized by increased integration with the world market. No longer does the demand for U.S. farm products depend just on domestic population and income growth, it also depends on the much less certain levels of foreign demands and foreign crop production. As the preeminent exporter of wheat, wheat flour, coarse grains, and soybeans, the United States both influences and is influenced by world markets and world events to an extent unprecedented since the United States' colonial years.

Partly because of these emerging interdependencies, this new era also could be characterized as an era of uncertainty. The experts' images of the world's future outlook for food vary widely, as do their projections of future demands on U.S. agriculture. The images can vary from an optimistic scenario of abundant and low cost food (Kahn, 1976) to that of a Malthusian doomsday trap characterized by widespread starvation (Hardin, 1974). The most pessimistically inclined see food scarcity as a permanent feature of society. One modern day Malthusian, Garrett Hardin, argues that there are not "shortages of food," but "longages of people," and sees starvation as the only method of restraining third-world population growth (Hardin, 1979). A contrasting view is voiced by those individuals who feel that agriculture has "an inherent and chronic capacity of overproduction" (Quance, 1976). Both extreme views can lead to a complacency with respect to the world's ability to feed itself. Acceptance of Malthusian thinking results in no policy recommendations other than inhumane indifference toward the poor. Belief in the chronic abundance hypothesis means there is no shortage problem, although there may be severe food distribution problems.

Between the optimistic and pessimistic extremes, of course, are various opinions that facilitate policy prescriptions for food-supply management. Lester Brown, for instance, raises doubts as to "whether it will be possible to get a combination of cropland expansion and yield increases that will satisfy the growth in world demand projected for the remainder of this century" (Brown, 1979, p. 14). He argues that an era of relatively abundant land has now come to an end, and that policies should be directed toward protecting the agricultural land resource. Robert Sugerman agrees with Brown that this is an era of resource limits but suggests that there must be considerable effort expended in developing the appropriate institutions for resource management (Sugerman, 1980). Keith Campbell, who is highly optimistic about world food production prospects, also assigns an active role to public policy, but he emphasizes the role of publicly supported research (Campbell, 1979).

The various policy prescriptions stem in part from alternative views of future food supplies. Cochrane (1979) succinctly describes this broad range of opinion: "As of 1978, there was no consensus among the experts with respect to the probable long-run food and agricultural situation in the world."

Within the context of this general uncertainty, numerous projections have been made of the future demand for U.S. farm products. It is generally agreed that U.S. domestic food consumption will rise only modestly (the FAO [1979] projects 0.9 percent per year) in the years ahead. Virtually all experts believe that the United States will have little trouble meeting this domestic demand. The important prospects for demand growth—and the most highly variable forecasts—are those for exports.

The FAO projects that, even under optimistic assumptions about the performance of their domestic farm sector, 90 developing countries might raise their net imports of grain from 33 million metric tons in 1975 to 52 million in 1990 and 88 million tons in 2000 (FAO, 1979). The International Food Policy Research

Institute (IFPRI) estimates that by 1990 developing countries will need to triple food imports unless they can fundamentally improve their own production capability (IFPRI, 1977). As the world's leading grain exporter, the United States could be called on to supply much of this grain.

Other projections are more specific about U.S. export prospects. The World Bank projects that U.S. wheat exports will rise by about 1/3 by 1990, while corn exports will rise only marginally (Bale, 1980). Abel projects, assuming that the real prices of U.S. farm products do not rise, that the U.S. could approximately double its exports of wheat, rice, feedgrains, and oilseeds over the next 25 years (Abel, 1980). The USDA, using a multiple scenario approach, estimates 1985 net grain exports of developed countries, chief among them the United States, ranging from 59 percent above to 16 percent below their 1973-76 average (USDA, 1978).

Whether U.S. agricultural exports will grow depends on several factors. First is the question of whether developing countries will be able to afford continued upgrading of their diet, particularly an increase in meat consumption. Will they be successful in raising domestic food production? If not, will they be able to export enough of their own raw materials and manufacured goods to generate the foreign exchange needed to buy U.S. food? Another consideration is the role of international protectionism, which could erect further barriers to trade in agricultural commodities. This applies particularly to future policy on food imports by the European Economic Community and Japan. These countries have the capability of increasing the local share of the food they consume, although it would be costly for them to do so (Bale and Greenshields, 1978).

A third factor is the potential for output increases by exporters whose agricultural products compete with those of the United States. These include traditional producers who have the possibility of increasing acreage or productivity, such as Australia, Canada, and Argentina for wheat. They also might include countries that have newly become important exporters, such as Thailand for coarse grains and Brazil for soybeans.

A final consideration is the price at which the United States offers its agricultural exports. In past years, the United States has been able to increase supplies greatly to the export market at little or no long-term increase in real prices. Foreign dependency on U.S. produced food has been encouraged by low prices; rising real prices would both dampen demand and encourage more production elsewhere. An important question for predicting the future, then, is whether the United States faces physical or economic barriers to continuing the exporting of major agricultural commodities.

AMERICAN AGRICULTURAL PRODUCTIVITY: LESSONS FROM HISTORY

If the past provides a guide to the future, there will be ample supplies of low cost U.S. agricultural products. Indeed, one author has stated in a related context that:

> Those who are most easily depressed about the precarious future of Western civilization are usually people who do not know the full history of its past [Highet, 1954].

Although such a philosophy may be overly optimistic, it is instructive to review American agricultural history in order to gain insights into the future. American agriculture has undergone considerable change as it developed into a highly productive industry. Identifying the nature and extent of the forces that determined the location, type, and productivity of past agricultural enterprises can aid us in analyzing whether or not future food and fiber will be available without substantially increased costs. Although all factors will continue to influence American agriculture, some will have a more dominant role than others. By understanding the past sources of American agricultural productivity, the future can be predicted with more accuracy.

For example, there have been at least two periods in which agriculture faced barriers to expansion. The first was the apparent constraint of a limited labor supply in the early 1800s. During this period, the cultivation of 100 bushels of wheat on five acres using broadcast seeding, walking plows, and brush harrows took between 250 to 300 hours of labor. Farm size was limited in the nonslave states by the amount of acreage a family could operate efficiently.

As farmers sought means to increase the harvest per worker, they developed and adopted numerous mechanical aids. By the 1850s, farming was undergoing a mechanical revolution. Plows, harrows, planters, cultivators, reapers, and threshers were rapidly accepted and used.

The second ostensible barrier to the expansion of agriculture accompanied the closing of the land frontier in the early 1900s. "Land, that seemingly inexhaustible resource, was in fact exhausted . . . . The number of crop acres harvested increased . . . but these acres were low production, marginal acres . . ." (Cochrane, p. 110). Indeed, the concern over the amount of cropland available for the future was reflected in a 1923 report on the use of U.S. lands (USDA Yearbook of Agriculture, 1923). This report estimated that, to support an additional 44 million people, 40 million more acres of cropland and improved pasture would be required in addition to half the acreage previously devoted to production for the export market. But the land constraint was eluded, too, first by increased irrigation and then by applications of knowledge developed by colleges of agriculture on plant and animal genetics, nutrition, and disease control that resulted in dramatic increases in agricultural productivity.

A review of agricultural history also will reveal that American agriculture has changed considerably; it has, in general, displayed increased specialization at the expense of regional self-sufficiency. Only in the nation's early history were its agricultural enterprises dispersed such that each geographical region supplied most of its own needs. Today, commercial agriculture is specialized by region and is concentrated; 10 states account for most of the value of agricultural products sold. Over 50 percent of the total value of U.S. agricultural products comes from the top 10 producing states of California, Iowa, Texas, Illinois, Nebraska, Kansas, Minnesota, Indiana, Wisconsin, and Missouri. As a result, crops and livestock must be transported long distances from farm to market.

Major crops have changed in location and importance. Cotton is the obvious example. Cotton, which at one point dominated the export markets, once stretched across the southeast; this region now grows soybeans and grazes cattle on former cotton land. Hay and oats were once more extensive than they are now, in part because of the need to feed the horses and mules that predated machinery power. In 1920, for example, there were over 17.2 million horses and 4.6 million mules in America.

Today, of course, farmers are not relying on horsepower. In 1977, for instance, farmers were using 4.4 million tractors (representing 232 million horsepower), 535,000 grain combines, 605,000 cornpickers, and 615,000 pick-up balers. Farmers are using purchased inputs of feed, seed, fertilizers, insecticides, herbicides, and veterinary supplies to a great extent. Because of these technological advances, farmers can now produce 100 bushels of wheat on only three acres with a tractor, 30-foot sweep disk, 27-foot seed drill, 22-foot self-propelled combines and trucks, with only three to four hours of labor.

These advances also have enabled significant growth in the size of the average farm, which is now over 450 acres. The current trend is toward larger but fewer numbers of farms. The total number of farms in 1979, for example, was 2.3 million, down from the 2.7 million counted in the 1969 census, the 5.4 million counted in the 1950 census, and the 6.5 million counted in the 1920 census.

Forces Determining Past Agricultural Development

The reasons for the changing trends in agricultural development are many and complex. First, there are the natural characteristics of the land: topography, rainfall, growing season, and soil productivity. There are also the characteristics of plants and animals that allow them to flourish in some areas but not in others. In addition, transportation improvement and the invention of refrigeration provided the stimuli for many regional shifts of agricultural product locations. Changing relative input prices, increased mechanization, improved production technologies, and changing consumer demands, including trade patterns, all influence crop and livestock location and modes of production. Finally, the social and political institutions of each historical time period were and are major factors influencing agricultural trends.

Physical Characteristics.   Among the most obvious influences on agricultural development are the physical attributes of commercial animals and plants, and the lands on which they reside. For example, the Corn Belt of the United States is one of the most productive areas of the world. The land is level or gently rolling, with rich soils. Sufficient and well-distributed rainfall, hot summer days, and warm nights make it ideal for growing grains, grasses, and legumes. "In fact, of the 100 million acres in the United States that soil technicians describe as excellent for grains, grasses, and legumes, about 3/4 are in the Corn Belt" (USDA Yearbook of Agriculture 1958, p. 122). The cropping systems used are such that they produce a large supply of feeds, both for local use (especially for hogs) and for export. In contrast, the Northeast has many acres that are rough, stony, and difficult to plow. Dairy cows, however, can make good use of such lands.

Thus, physical characteristics of land, plants, livestock, and their products can, to some extent, determine the location of agricultural enterprises. While it is true that one can grow oranges in Alaska—at a high cost, heated greenhouses could make Alaskan orange orchards quite productive—there are many economies to be gained if the crop is matched to a region with a suitable soil and climate.

Transportation Improvements. Another important factor influencing method, type, scale, and location of agriculture is the cost of transportation. The Northeast region of the United States was at one time largely self-sufficient in most commodities. In 1850, for example, it had many vegetable and fruit enterprises. With the advent of both low-cost transportation and refrigeration, the Northeast steadily has reduced the number of its acres in fruits and vegetables and instead has relied on shipments from California, Texas, and Florida. The cheaper imported produce meant that many farms in the Northeast were no longer profitable. Indeed, total acreage in farms declined over 50 percent from a World War II peak to the 1970s, and much of the land reverted to trees.

Similarly, access to feed is very important in poultry production; feed accounts for 2/3 to 3/4 of the per pound cost of a chicken before slaughter and of the cost per dozen eggs. Poultry enterprises are therefore located in such a way as to reduce feed costs—for example, in places where a canal and river system enable low-cost barge shipment of grain or where producers can take advantage of local feed production.

Changing Relative Input Prices. Until 1940, commercial agriculture was mainly an extensive agriculture, that is, an agriculture that expanded its production not by improving yields but rather by expanding cultivated acreage. One reason for this type of early agricultural expansion was simply that land was inexpensive relative to labor and capital. In fact, changes in the relative input prices underlie much of the interregional shifts that have taken place in agricultural production throughout our history. One of the most important factors influencing the organization of agriculture in the West, for instance, was inexpensive water for irrigation. Similarly, early experiments with the planting of soybeans in the South in the late 1930s were partially a result of increased labor costs. "'The average labor requirements for producing an acre of cotton in the Louisiana Mississippi Delta area,' pointed out a Louisiana State University bulletin in 1943, 'are 183.6 hours; . . . and an acre of soybeans, 9.6 hours'" (Fornari, 1979, p. 251).

Improved Mechanization and Production Technologies. Improved mechanization and production technologies also have considerably influenced agriculture. Farmers have increasingly relied on inputs that are products of incredible technological advances. These developments—automated farm machinery, fertilizers, fungicides, pesticides, herbicides, and antibiotics—made intensive production of single crops feasible. Furthermore, as farmers adopted these technologies, interregional shifts in crop production sometimes resulted. Invention of the steel plow in the mid-1800s, for example, opened the thick, sticky soils of the prairies to cultivation. Soybeans and cattle moved into areas of the south formerly devoted to cotton as cotton growers sought reduced costs:

> A most important factor affecting increased soybean cultivation was the continuing shift of cotton growing to newer and larger areas in the West and Southwest where the terrain and climate were conducive to larger scale mechanization and where the use of irrigation reduced the risk of crop failure and greatly enhanced production possibilities" [Fornari, p. 251].

The use of the gasoline engine meant that vast acreages formerly devoted to hay and oats production for animal feed could be released to other commodities. As farm equipment grew larger and embodied more horsepower, farm size increased.

It took America several decades to develop hybrid corn, but the results were spectacular. Careful genetic selection produced corn varieties that increased yields as much as three times (USDA Yearbook of Agriculture, 1975). Antibiotics revolutionized the livestock and poultry industries.

Changing Consumer Demands. Another factor of considerable importance in agricultural development is that of changing consumer demands. World trade, for example, spurred the demand for soybean meal and soybean oil. When this occurred simultaneously with the increased production of synthetic fibers, King Cotton was dethroned. The changes were dramatic. Whereas in 1951 synthetic fibers had only 5 percent of the fiber market, a quarter of a century later they had a 68 percent share of the market (Fornari, 1979). The growth in soybean cultivation was equally dramatic. In 1950, 14 million acres of soybeans were harvested in the United States; in 1978, 63 million acres were harvested.

Similarly, the strong increase in demand for grain-fed beef does much to explain structural and location changes in commercial feeding operations. Per capita consumption of beef increased from 63 pounds in 1950 to 120 pounds in 1978; simultaneously, population increased from 151 million to nearly 220 million persons. This tremendous increase in demand translated into increased demands for feed grains as well. In addition, forage supplies were developed on acreage formerly devoted to cotton.

While consumers were demanding beef, they were reducing consumption of pork. As pork production declined, farmers found that agricultural land would yield more if planted in crops than if used for extensive hog pasturing. As a result, hog production shifted from extensive to intensive land use, with specialized hog buildings and equipment. However, the advantages of maintaining hog enterprises near the production of feedgrains discouraged any major interregional shifts. The Corn Belt still produces 2/3 of the total liveweight production of hogs.

Increased demands for U.S. agricultural products by foreign nations also considerably influence U.S. production and promise to be one of the major influences in the future. As one recent article noted, ". . . after the introduction of the tractor, the most important shock affecting the structure of American agriculture in this century has come from abroad in the form of increased market interdependence" (Carter and Johnson, 1979, p. 3). One of the most obvious impacts of our increasing exports is the increased production of feed grains in the United States in order to meet world demands; the volume produced is more than twice that harvested in the 1950s. U.S. cattle feeders are now in competition with foreign buyers for the purchase of feeds. This competition influences the price of feed and meat and the location of cattle herds.

Social/Political Infrastructure. A final major factor influencing location, type, and method of U.S. agricultural production is the social and political infrastructure. This can include government policies such as taxes, regulations, land-disposal policies, water-resource projects, commodity programs, and trade polices.

Commodity programs, in particular, can influence significantly the structure of production. For example, "present regulations insulate Northeast dairy producers from competition of producers in other regions, especially the Lake States" (Schertz, 1979). This has led to higher farm incomes in the Northeast than would otherwise be the case and has slowed the decline in numbers of Northeast farms. The tying of commodity price support programs to acreages and not production quotas encouraged farmers to seek methods to increase yields per acre. Three government programs—the Federal-Aid Highway Act, the water-resource development programs, and the rural electrification projects— probably did as much to influence agricultural production practices, methods, and locations in the twentieth century as did many of the other factors combined.

Other institutional influences on the farm sector are the inflationary and deflationary forces in the economy, and the credit policies of the banking sector. The desire to hedge against inflation, for instance, has been a factor in increasing the demand for land and equipment, and thus encouraging the consolidation of land into larger operating units. Credit expansion since World War II has enabled farmers to finance farm expansion.

Social institutions also have significantly influenced American agriculture's productivity. In the beginning of this nation's history, the adherence to a Jeffersonian view of property established the ideal of widespread private ownership of agricultural lands. This also led to land policies of the nineteenth century that emphasized the removal of lands from the public domain into private hands and provided the impetus for rapid development of all the nation's agricultural lands.

Accompanying this philosophy was another of considerable influence, a belief in free and broadly available education. Not only did this produce an educated farm populace, but also a strong growth in collegiate agricultural education and research that has continued through most of the twentieth century. In real terms, "resources devoted to agricultural research increased by more than five times between 1915 and 1970" (Cochrane, 1979, p.246). Private firms also engaged in research, concentrating their attention on developing technologies that made farmers even more dependent on manufactured inputs.

Along with research, colleges also developed a sophisticated extension program that served as a conduit of new agricultural knowledge. The philosophy of rapid dissemination of new technologies to an already well-educated farm populace did much to influence the course of development of American agriculture.

Interaction of Factors. The factors influencing the development of American agriculture have not operated independently. Physical suitablity for production, for instance, is not a static classification. Improved corn hybrids allowed corn production to migrate north and south of the formerly "suitable" corn producing region. Improved cattle breeds were developed for southern climates where cattle prosper on genetically improved grasses. Irrigation with inexpensive water vastly expanded the commercial use of the Great Plains, where some of the richest soils in the world lie in areas of low natural rainfall. And institutions such as crop insurance made it possible for farmers to plant corn in more marginal areas by reducing the risk to profits of a crop failure in areas where irrigation is not used. These same risk-reducing mechanisms allowed the farmer to specialize in one product by making diversification of the farm unnecessary. Commodity programs that supported prices of various agricultural products made some farming enterprises (e.g., small tobacco farms) profitable when they otherwise would not have been.

Given the interaction of all these forces, America simultaneously could develop, say, hybrid corn and introduce it to farmers who had the education, capital, tenure security, and financing to risk planting it on their acreage to produce the increases in productivity that characterize American agriculture's past.

AMERICAN AGRICULTURE: THE FRAGILE SECTOR?

The brief review of U.S. agricultural history illustrates that past harvest increases were influenced by numerous factors, including research advances, inexpensive and abundant land, water, energy, and good climate. Past agricultural research advances have allowed the development of an intensive and monocultured agriculture. Antibiotics and selective breeding have enabled farmers to concentrate livestock in small areas. Hybrid development and pesticides have allowed farmers to specialize in single crops so that U.S. agriculture is now regionally specialized. Relatively inexpensive transportation has further encouraged specialization, and food is grown far from final markets. Irrigation has opened millions of previously arid acres to cultivation. The existence of rich and fertile lands has reduced the urgency of soil conservation or the need for heavy fertilizer use. Furthermore, research results produced mechanical, biological, and chemical innovations that enabled agriculture to evade resource barriers.

Now, as the 1980s commence, there are those who think new constraints will become evident. One reason for this belief is that there has been a plateauing of yields of the major food crops: wheat, sorghum, soybeans, and potatoes [Wittwer, p. 69]. The reasons for this plateauing are many but the most often cited are that the most productive lands are already being used, productivity growth appears to be slowing, rising energy costs are constraining irrigation and the use of energy-intensive fertilizers, expanding production is exacting costs of soil erosion and compaction, overpumping of groundwater is reducing the availabilty of water, climate fluctuations and air pollution are reducing yields, and declining agricultural research support is inhibiting technological innovations [Wittwer, p. 69].

Indeed, it may be that some of the past reliance on the inputs that were relatively cheap—energy, water, land—when used on a high-yielding, highly mechanized monocultural agriculture is the source of increased vulnerability in a rapidly changing world. As one author states:

> . . . the winds of change can blow swiftly across agriculture. Food abundance is based on our great natural resource but has become increasingly unnatural as greater energy, chemical fertilizers, pesticides and irrigation water inputs are used in increasingly concentrated and monocultured production processes very much in oppositon to natural ecosystems [Quance].

If this is so, the need for new policy directions is evident. As Lee states:

> Thus, within the first half of the 1980s, the long period of adjustment and disequilibrium in U.S. agriculture, with all its attendant problems (and associated policies, programs and institutions) may phase into a new era of limits with all its attendant problems. Should that happen, the policies, programs and institutions designed to address the problems associated with chronic surpluses and disequilibrium would likely not be appropriate. In that case, the challenge before us is clear [Lee, p. 16].

The Concerns

The principal concerns with respect to the future of American agriculture are that (1) the U.S. agricultural resource base no longer has excess capacity, (2) technological advances may not be able to compensate for reduced productivity from such problems as soil erosion, (3) relative price increases of previously inexpensive inputs will seriously alter the profitability of current production practices, and (4) this trend, when combined with other factors, such as monocultural production or climate change, may make U.S. agriculture increasingly vulnerable to future shocks.

The concern that the U.S. agricultural resource base no longer has excess capacity reflects a belief that the amount of remaining rural land suitable for conversions to cropland is small and is being further reduced by conversions of cropland to nonfarm uses (USDA, Soil Conservation Service, 1979). Some of the very physical characteristics that make cropland fertile, such as level ground and rich well-draining soil, also make the same land desirable for development. The better agricultural lands often are selected for housing, highways, and shopping centers. Similarly, when dams are constructed, the reservoirs frequently inundate fertile land in the former floodplain. Once removed from the agricultural system, these lands can usually be returned only at a high cost.

The loss of agricultural land is considered more serious because of perceived stresses that reduce soil productivity. Erosion, for instance, is a natural process; if erosion rates do not exceed the formation of new soil, there is little or no reduction in soil productivity. However, when soil erosion greatly exceeds soil formation, as is frequently the case with intensive agricultural practices, part of the nations's cropland fertility literally is washed or blown out of the agricultural system.

In the past few decades, changes in soil productivity have been compensated by increasing yields due to technological advances in crop genetics and the use of fertilizers. But because of fears that future technological advances will not be as effective or as low-cost as in the past, soil productivity issues are becoming increasingly important. The "quality of cropland" concerns added to the "quantity of cropland" concerns cause many to raise doubts about long-term food security (Brown, 1979; Sampson, 1979).

Other characteristics of our present agriculture system, when considered in conjunction with land quantity and quality changes, raise the possibility of increased vulnerability of U.S. agriculture to future shocks such as water and energy shortages, physical-biological disasters, extreme weather variations, or changes in weather patterns.

U.S. agriculture has evolved to a point at which farmers are not as able to control costs as they once were, for there is an increased dependence on purchased inputs.

> In the days of horse and mule farming, in the United States, farmers' costs were generated within the farm economy. Feeds and fertilizer were produced on the farm. Insecticides and pesticides were fewer and less expensive. Power was provided by men and animals. Farmers had considerable control over these "inputs" and their attendant costs.
>
> Today, farmers are dependent on national credit systems. They rely on steelworkers, truckers, dockworkers. They depend on international sources such as the Organization of Petroleum Exporting Countries (OPEC). And this growing dependence on others is not exclusive with American farmers but is important in all countries. It has created an element of instability in the global farm economy that will focus increased attention on the policies of government [Bergland, 1979].

As shortages occur in some purchased inputs, such as oil or irragation water, farmers are forced to adjust to increased prices and/or reduced availability of these inputs. This adjustment will take place in the method, type, scale, or location of production and ultimately in the cost of food.

Further, to achieve high yields, a uniform product, and low production costs, crops have literally been "standardized." The fact that a single cytoplasm was used to produce nearly all hybrid corn grown in the United States made it possible for a single parasite mold, *Helminthosporium maydis*, to reduce the 1970 corn crop by 15 percent (USDA Yearbook of Agriculture, 1975). Other crops such as soybeans, wheat, and potatoes are also built on a narrow and therefore vulnerable genetic base. This reliance on monocultured crops increases the risk that vast acreages could be subjected to the same disease, pest infestation, or climatic change. The stability that comes from ecological diversity has been reduced, perhaps significantly.

Finally, there is concern that climate variations could significantly change crop yields or even crop locations. If U.S. agricultural productivity has partly been the result of abnormally favorable weather conditions, as some climatologists think, then there will be adverse effects on food production if the climate reverts to a more "normal" pattern. There is also concern that agriculture may be unable to absorb easily the shocks of a sudden change in the weather patterns. Indeed, the recent volcanic eruption of Washington State's Mount St. Helens is cause for reflection. When Mount Tamboro in the Dutch East Indies erupted in 1816 it was followed by "the year without a summer." New England had killing frosts in July and August of 1980; earlier, Europe had substantial crop failures (*Newsweek*, June 2, 1980). The U.S. agriculture system is obviously not immune to such natural disasters.

The Agricultural Resource Base

Does U.S. agriculture have the resiliency to absorb these types of "shocks" without severe dislocations or large increases in food prices? Are the concerns discussed here valid? These issues are the subject of the chapters that follow.

Competition for Agricultural Land. Whether or not the agricultural resource base is shrinking due to increased demands for agricultural lands by alternate uses such as highways, reservoirs, urbanization, recreation, or energy production is the subject of Philip Raup's paper. He notes that, on a nationally aggregated basis, acreage used for crops has been, with few exceptions, relatively stable from 1910 until 1978. However, the national data mask many changes in regional uses. The Northeast, Southeast, Appalachia, and Southern Plains have lost land in agriculture; all other regions have gained. Generally, cropland losses have been timberland gains, since trees have grown up on lands abandoned as unprofitable for farming. Raup examines other alternative uses for land and concludes that for some uses—highways and reservoirs—historical trends do not provide good guides to the future. He states: "The competition for land that was fostered by the boom in highway construction is still with us. It will be some years before the echo-effects have been assimilated in land-use patterns. But it seems reasonable to conclude the major effects are behind us." Similarly, "our dam-building era, like our highway-building era, is largely behind us." Raup proceeds to point out that a new form of land use is emerging, which he terms "agri-urban." Agri-urban is characterized by an intermixture of farm and rural residential uses with no clearly defined boundaries for either use. This pattern, when coupled with increased housing demands and demands for recreational areas, suggests a major influence shaping the future competition for rural land.

Land used for energy production, such as coal strip-mining or ethanol production, is probably not going to place large demands on agricultural land, according to Raup, unless fuel development is highly subsidized. Otto Doering's paper reinforces this opinion; he maintains that energy farms with special energy crops are probably not feasible on a large scale and "food and fiber crops appear to have the edge over energy crops in the competition for land."

Raup then places these alternative uses in perspective by discussing factors that influence interregional competition for land. He skillfully traces the impacts that increased irrigation patterns have had in causing regional shifts for corn, sorghum, alfalfa, and beef feeding. Now, states Raup, because of increased world demands for beef, competition for land in the United States has entered an international phase where the "most acute competition for land in the United States today is between foreign and domestic producers of meat." This has led to a restructuring of competition for land, the full effects of which are not yet evident. Indeed, Raup asserts that the export market is now the most important factor affecting interregional competition for land within the United States.

Two factors, then, emerge from Raup's analysis as strong influences on future competition for aricultural land: urbanization and foreign trade. The first may put less pressure on agricultural lands conversion than might be supposed, due to the rising real costs of credit and energy. However, increasing acreages of agri-urban land may restrict the choice of size and intensity of farm enterprises. The second factor—foreign trade—will not only provide impetus for interregional shifts of production but could conceivably call for expanded agricultural production on existing lands and the opening of new lands to agricultural production.

Soil Productivity. Frederick Swader's paper addresses a growing concern: whether or not the nation is "mining" a resource unwisely and sacrificing future productivity for present yields. Soil scientists have established a "tolerable level" or "T-value" for various soil types. This is the level of erosion that can be sustained without reducing soil productivity. In the Great Plains, Corn Belt, and Delta States, over 77 million acres exceed these limits. The USDA has projected how maximum yields in the year 2030 would be affected by the assumed continuation of present erosion losses. USDA concluded that the greatest declines in productivity would be in soil groups IV and V, land capability classes that generally are less suited to cultivation. Yield differences on these soils due to erosion were as much as 46 percent. If foreign-trade pressures cause more cultivation of these less suitable lands, reduction in overall yields can be substantial. If, however, in the future, agriculture uses fewer acres than at present, presumably many of these less suitable acres will be retired from production. Reductions in yields by erosion from the soil group classes I and II, in contrast, ranged from zero to 3 percent. While much more reassuring than a 46 percent difference, a 3 percent decline in today's production of wheat alone, for example, would equal 60 million bushels, worth $240 million annually at today's prices. High export demands would accelerate these losses if increased use of lower-quality soils was necessary. Indeed, one study estimated that under a high export scenario, increased soil losses due to erosion were likely to range from 40 to 106 percent in the Corn Belt states.

Swader also carefully examines other factors that may be putting stress on agricultural productivity, such as soil compaction or a decline in the use of fertilizers. At least one study suggests that crop losses to soil compaction are significant, amounting to over $1.18 billion in 1971. In current dollars, these figures suggest losses of over $3 billion a year. Furthermore, U.S. agricultural productivity is heavily dependent on chemical fertilizers, and "there is a consensus that it would be virtually impossible for the United States to remain an agricultural exporter without chemical fertilizers." But Otto Doering points out that, although the per acre energy content in gasoline equivalents for fertilizer is nearly 40 gallons, there does not appear to be a realistic expectation of energy shortages strong enough to induce farmers to adopt cropping systems (e.g., legume rotations) that provide much of their own nitrogen at a cost of less output of traditional grain crops.

Kenneth Frederick's paper analyzes another stress placed on agricultural productivity: increased soil and water salinity resulting from increased irrigation. He suggests that 25 to 35 percent of the irrigated lands in the West have a salinity problem. "Undoubtedly," he says, "there will be some resulting decline in productivity and profitablity."

Research and Productivity

In the past, the United States has been able to increase productivity per acre despite ostensible natural and physical constraints. For example, any losses in soil fertility due to past erosion have been masked by increases in yields per acre due to improved crops and the application of chemical fertilizers. Vernon Ruttan addresses in his paper the contribution of research to agricultural production and the future of agricultural productivity gains. After reviewing many research studies that demonstrate the very high past returns to research in agriculture, he focuses on whether or not growth in productivity can be sustained. Ruttan is skeptical for several reasons. First, government research funding has been declining; second, there will continue to be a lag until we see new technical changes induced by rising agricultural energy prices. Ruttan suspects that productivity growth will be more in line with the 1925-1950 experience than with the 1950-1980 experience. That is, he anticipates productivity growth closer to 1 percent per year than to the 2 percent per year associated with the last three decades. If Ruttan is correct, the effects of losses in soil quality and soil productivity will be more noticeable in the future. This is particularly true if agriculture continues to expand by harvesting new lands (as opposed to more intensive farming of existing lands) that, in many cases, are less productive and more conducive to soil erosion losses than the presently harvested areas.

Relative Price Increases

Another concern analyzed by these papers is that changes in the relative prices of water and energy will alter the profitability of current production practices.

Water Supply. Frederick's paper comprehensively analyzes the U.S. water supply situation. He concludes that irrigation cannot be expected "to continue to contribute to agricultural expansion in the way it has in the past three to four decades." Much of the expansion in the past was due, in part, to governmental subsidies. Although this resulted in lower food prices, it also has meant higher taxes to pay for artificially "cheap" water and commodity support programs. In addition, it has meant declining groundwater tables that have resulted in increased pumping costs and the "premature" mining of our aquifers. Frederick's analysis suggests that, except for Nebraska, where some additional expansion is probable, long-term Western groundwater irrigation is likely to decline because of rising pumping costs.

Raup's paper corroborates this point, stating that these changes in water availability could dramatically shift the present geographic pattern of land use and greatly alter the nature of competition for land in the Midwest and Great Plains. Raup, noting that feed for cattle fed in the southern Great Plains has been grown with water from the Ogallala aquifer, states that "unpriced water . . . has been capitalized into a national level of beef consumption that cannot be sustained in the long run without a return to the feed grain supplies of the Corn Belt. We have a fed-beef economy that has become dangerously dependent on an exhaustible resource base." Frederick feels that irrigation will contribute to agricultural production in the future, but that it is imperative that water be priced to reflect its true scarcity. Water priced as if it were abundant, when it is not, will be used inefficiently.

Energy Dependence. Doering's paper examines agriculture's response to the rising price of energy. He emphasizes that energy inputs in agriculture have been used in the past not just to increase yields per acre but to reduce the risk of a harvest failure. When energy is a small proportion of total production costs and when the energy input is important to accomplishing the harvesting or processing in a timely and less-risky manner, farmers will be far less sensitive to price changes than when this is not the case. Ultimately, agriculture will adjust to the new relative price differences, but Doering believes that this adjustment probably will not include a substitution of labor for energy. Rather, changes will occur in cropping systems, in irrigation-dependent crops, in the trend toward the geographical concentration of certain commodities, and in the comparative cost of different commodities. While Doering suggests that agriculture can absorb these changes, he cautions that decisions to change, adapt, expand, or go out of business will be made at the farm level. Therefore, he argues, there is a strong case for not making choices about adjustments to production on a centralized basis. Furthermore, he predicts strong pressures to protect various farm groups from the price increases and hence from the need to adjust to changing prices and availability.

Crop Vulnerability

The final concern analyzed by these papers is that U.S. agriculture is vulnerable to the shocks of extensive insect and disease damage and climatic change.

Monoculture. Plant geneticist Jack Harlan considers the implications of U.S. agriculture's dependence on monoculture. Harlan emphasizes that developing a crop monoculture has both historic roots and an economic rationale. Although he concedes that in some crops, such as corn and sorghum, this means increased vulnerability to disease and pests, he suggests the risks can be managed by improving and evaluating national and international germplasm collections, by improving our understanding of nondomestic diseases and pests, by stronger plant breeding programs, and by maintaining competition among the various seed companies. Further suggestions by Harlan include encouraging tolerance of a higher degree of diversity within a product in the marketing and processing sectors, development of an effective professional management team to deal with epidemics, the development of safer agricultural chemicals, and the training of multidisciplinary scientists. Harlan states his personal optimism that the benefits of crop monoculture can continue without unacceptably high risks of devastating disease or pest infestations.

Climate Change. Finally, Robert Shaw discusses the difficult subject of climate change and agriculture's dependence on present weather patterns. His paper offers a striking statement: "Agriculture as we know it has developed during an abnormally warm period in recent climatic history." Then Shaw proceeds to argue: (1) "the climate of the future cannot be predicted," and (2) "for the present, we must be more concerned about year-to-year variation than about long-term trends." Furthermore, we are adding a considerable number of human-made variables to the climate system: carbon dioxide, ozone, heat, acid rain, atmospheric particulates, and intentional weather modifications. Each future climate scenario, predicted with great uncertainty, can be associated with future yields and agricultural production locations, also predicted with great uncertainty. A large cooling of the climate, for instance, might improve yields in the United States at the expense of yields in the U.S.S.R., Canada, and China. However, all long-term climate scenarios tested indicated little change in total world production. Shaw suggests, however, that a series of extreme

short-term climate variations may produce serious food shortages. He questions:

> What would be the result if a combination of events happens
> such that a major drought occurs in a large area of the world,
> at a time when no surplus grain is available in other parts of
> the world? The thought is unpleasant; the results of such an
> event would be much more unpleasant. Who would make the
> triage decision?

Other Factors Influencing the
Future of American Agriculture

There are, of course, other factors that may influence the future
of American agriculture. Two that come to mind readily are foreign ownership of American farms and environmental constraints
on present and future agricultural practices. While these topics
are not the subject of any of the papers in this book, they are
worthy of additional elaboration.

Foreign Ownership of Farmland. Beginning in the middle
1970s, considerable concern was expressed in Congress and the
news media over foreign purchases of U.S. farmland. A number
of very large purchases by foreigners were noted in several parts
of the country—for example, Italian investment in a 127,000-acre
Oregon ranch, German purchase of 1,000 acres of Iowa farmland,
and purchase of Arthur Godfrey's $5 million Virginia farm by a
Saudi Arabian. The fact that no accurate measure existed of the
number of acres actually owned by foreigners, or of the rate at
which that number was increasing, heightened fears that foreign
ownership was an important and rapidly spreading phenomenon.

Foreign purchases were attributed to such diverse causes as a
desire to protect capital in an unstable world, a depreciating dollar
that made U.S. land cheap in terms of foreign currencies, and tax
laws favoring foreigners over U.S. residents. Concern over the
effects of the purchases ranged from simple nationalism to more
sophisticated contentions that foreign landlords might not protect
soil from erosion nor contribute much to local economic development.

In 1978, Congress passed the "Agricultural Foreign Investment
Disclosure Act," which required foreign owners of one or more
acres of agricultural land, both cropland and forest, to report the
holdings and transactions to the U.S. Secretary of Agriculture.
The most recent report under the act reveals foreign ownership

of 5.6 million acres, or less than 1/2 of 1 percent of U.S. agricultural land (USDA, 1980). Reported foreign ownership amounted to more than 1 percent of the agricultural land in only four states— Maine (5.1 percent); Nevada (2.1 percent); South Carolina (1.5 percent); and Tennessee (1.3 percent). The high figure in Maine is due to a large holding of woodland by a single timber company. The largest foreign holdings were by U.S. corporations with ownership interests from the United Kingdom, Canada, Luxembourg, or West Germany. Individuals with addresses in Canada, West Germany, and the Netherlands Antilles (a popular tax haven) also reported significant ownership.

Forest land accounted for 46 percent of the foreign holdings, most of it land in the southern states held by U.S. timber companies with foreign interests. The report also found that between February 1979 and February 1980 foreigners made net acquisitions of 600,000 acres of U.S. agricultural land. Because farmland turnover is so slow, at current rates of purchase it would take nearly 20 years for foreigners to obtain even an additional 1 percent of farmland.

Even if foreign ownership were widespread, it is far from certain that it would be a threat to national security. Many foreigners are simply investors, managing the land in very much the same way as do nonresident U.S. investors. It also can be argued that some foreign ownership is an asset to our security, since it gives foreigners an economic stake in the prosperity of the U.S. economy. One former State Department official has argued in private conversation that the United States is more secure if foreigners hold title to farmland, which could be easily expropriated in time of crisis, than it would be if they held U.S. dollars or other liquid assets, which might be dumped suddenly on the market, conceivably creating a worldwide financial panic. In any case, given the small amount of foreign ownership discovered thus far, it is difficult to regard it as a significant threat to U.S. agricultural production.

Environmental Constraints. Yet another potential factor influencing U.S. food security is the possibility that production cannot be increased without severe damage to the environment. As an example, pesticides in many cases kill not only the target pest but beneficial species as well. DDT, which was found to have disastrous effects on the reproductive ability of some birds, is a case in point. Moreover, some pesticides and herbicides have been found to be carcinogenic or mutagenic, posing threats to human health. Because of health dangers, application of such widely used pesticides as DDT, Mirex, heptachlor, and chlordane has had to be banned or severely limited. Concern also has been raised about food residues of hormones and antibiotics, both used to promote animal growth.

Another environmental impact of agricultural expansion is damage to wetlands and other wildlife habitats. It is estimated that 51 million acres, or 40 percent of the original wetland area of the United States, have been converted to other uses thus far (Ladd, 1978). Even now, agricultural expansion continues to promote the draining of "pothole" wetlands in the Upper Midwest, "pocosin" wetlands in North Carolina, and bottomland hardwood swamps in the Mississippi Valley. All of these had sheltered or provided food for a variety of terrestrial and avian species of wildlife. As Frederick points out in his paper, irrigation expansion has already brought several western river systems to or below the water level needed to support "instream uses." One of those uses is habitat for fish and other water dependent species.

Yet another environmental effect of agriculture is sedimentation, the off-site result of soil erosion. The U.S. Soil Conservation Service (USDA, 1980) estimates 40 percent of current sediment load is produced by cropland. Sediment particles pile up behind dams, impede navigation, increase water treatment costs, injure fish populations, and provide a transport medium for toxic substances. If expansion of agriculture onto marginal lands produces more erosion, as Swader indicates, it is likely that stream sediment loadings also will increase. Agricultural runoff also produces other forms of water pollution, such as increases in phosphates (which can produce algae blooms) and the introduction of nitrates, herbicides, and pesticides into drinking water supplies.

In most cases, environmental factors do not completely constrain production but result in the production of more food at the expense of other important values. Some of the effects are avoidable through simple modifications of farming practices. Others are more difficult to prevent and may cause us to make social choices limiting some forms of agricultural expansion.

CONCLUSIONS

American agriculture is a strategic resource, and the maintenance of national security includes maintaining the long-term sustainability of agriculture. After considering whether or not there are serious threats to the future of American agriculture, the authors of the seven papers have fashioned positions we might describe as "guarded optimism." They have discounted some possible concerns, such as monoculture crop vulnerability and losing land to highways and reservoirs, but have introduced others, such as increased salinization of western soils and competition for land from subsidized energy crop production. Collectively, the papers give pause.

Perhaps the most disquieting paper is that of Vernon Ruttan, for our brief examination of U.S. agricultural history has demonstrated that increased productivity has enabled U.S. agricultural production to expand without pressing against severe physical constraints. As Ruttan indicates, the potential for expanding agricultural lands by bringing new lands under cultivation was ending near the beginning of the twentieth century; it was literally the "end of the frontier." Agricultural production growth then became a product of increased mechanization and fertilization. These resulted in high annual rates of labor productivity growth as other inputs were substituted for labor. But, until a scientific revolution took place, there was relatively slow growth in output per unit of *all* inputs. Changes in relative prices eventually induced advances in biological as well as mechanical technology. If Ruttan is right, and future productivity increases are more in line with the slow growth era of 1925-1950, and if increased demands for U.S. agricultural products do indeed materialize, then the United States may witness a return to an agriculture in which increased yields depend mainly on the use of more agricultural inputs.

As Ruttan states:

> The closest analogy to the present situation in American agricultural history was the period between 1900 and 1925. With the closing of the frontier, productivity growth declined. The new sources of productivity growth, chemical and biological technology, did not begin to emerge for several decades. My own guess is that it will be at least another decade before the direction of technical change . . . becomes clear.

While America is awaiting the hoped-for new technical change, there may be a frustrating period of adjustments to constraints. Three inputs—land, water, and energy—will probably have rising real prices in the future. Several of the authors have concluded that this in turn will mean increasing pressures for interregional shifts in crop and animal production as well as changes in production practices. Increased irrigation costs and rising demands for U.S. grain may, for example, move the livestock industry back toward the grazing-based locational patterns of the nineteenth century. If this happens, cattle herds will put greater demands on the land base than ever before, and upward pressures on prices are the obvious outcome. For another example, increased transportation costs may mean local vegetable produce can compete in Northeastern urban markets with imports from distant parts of the country.

The uncertainties of land-use patterns resulting from the new demographics, new trade patterns, and possible new land uses, such as the subsidized use of agricultural lands for energy production, compound the problems of predicting the nature of interregional shifts and the magnitude of any pressures on input and product prices. Fully apparent, however, is that the present prices paid by consumers, both domestic and foreign, do not reflect the true long-run social costs of that production. Soil erosion, soil compaction, salinization, declining aquifers, water pollution, and loss of wildlife habitat are not reflected in the market price of food. Exactly how these nonpriced costs will eventually affect the future productivity of this nation's agricultural resource is uncertain.

These uncertainties cause some of the authors to suggest caution; yet none are prone to "sound the alarm." Predictions based on past and present trends do not suggest anything approaching a crisis or catastrophe. But these predictions describe the authors' opinion of the occurrence of the most probable events, not the occurrence of less likely, but possible events. Perhaps American agricultural policy should be more sharply focused on insuring against extreme deviations from the most probable event. The most obvious example of such a deviation is a climate change to one less favorable to U.S. agriculture. The amount of insurance that might be appropriate would, of course, depend on the costs and benefits (and the distribution of the costs and benefits) of the insurance strategy selected. The insurance possibilities include: wide spatial distribution of agricultural lands, more diversity in agricultural production by region, more agricultural research funding, private and public grain reserves, programs to reduce specific "threats" to agriculture, modifying government programs that encourage a structure of agriculture vulnerable to extreme events, or pricing agricultural products to reflect all private and social costs.

Needed research relevant to these strategies would include determining: the future costs of soil erosion, compaction, and salinization; the impacts on U.S. agricultural productivity of a severe temporary or longer-run climate change; the costs of converting noncropland to cropland; and the trade-offs in productivity resulting from wider and more diverse spatial distribution of agricultural enterprises. In most situations, it will be easier to identify the costs incurred when implementing an insurance strategy than it will be to estimate the benefits from protecting against an extreme event.

Policy debates should not await research findings, however, if for no other reason than that policy debates can help us identify our areas of ignorance. With the present farm economy characterized by surpluses and credit and cost squeezes on the farmer, it may seem strange to suggest debating how we might confront possible long-run shortages. However, the urgency of near-term problems must not obscure the importance of considering how to organize an agriculture sustainable for as long as we expect our society and economy to endure.

REFERENCES

Abel, Martin E. "Growth in Demand for U.S. Crop and Animal Production by 2005." Paper presented to RFF Conference on the Adequacy of Agricultural Land, Washington, D.C. 19-20 June 1980.

Bale, Malcolm. "Market Prospects for Grains." Unpublished World Bank Staff Paper, January 1980.

Bale, Malcolm D., and Greenshields, Bruce L. "Japanese Agricultural Distortions and Their Welfare Value." *American Journal of Agricultural Economics* 60(1978): 59-64.

Bergland, Bob. "Attacking the Problem of World Hunger. *National Forum* 69(1979):3-6.

Brown, Lester R. "Population, Cropland and Food Prices." *National Forum* 69(1979):11-16.

Butz, Earl L. *Food Power: a Major Weapon.* Speech, Washington, D.C., 24 June 1974.

Campbell, Keith O. *Food for the Future.* Lincoln: University of Nebraska Press, 1979.

Carter, Harold O., and Johnson, Warren E. "Some Forces Affecting the Changing Structure, Organization, and Control of American Agriculture." *American Journal of Agricultural Economics* 60(1978):738-48.

Cochrane, Willard W. *The Development of American Agriculture: A Historical Analysis.* Minneapolis: University of Minnesota Press, 1979.

Dennison, Edward F. *Accounting for Slower Economic Growth: The U.S. in the Decade for the 1970s*. Washington, D.C.: Brookings Institution, 1979.

Food and Agriculture Organization, United Nations. *Review of Food Consumption Surveys*. Rome: FAO, 1979. And *Agriculture: Toward 2000*. Rome: FAO, 1979.

Fornari Harry D. "The Big Change: Cotton to Soybeans." *Agricultural History* 53(1979):245-68.

Gilmore, Richard. "Grain in the Bank." *Foreign Policy* 38(1980):168-81.

Hardin, Garrett. "Living on a Lifeboat." *Bio Science* 24(1974):561-68.

Hardin, Garret. "Can Americans Be Well Nourished in a Starving World?" *National Forum* 69(1979):25-27.

Highet, Gilbert. *Man's Unconquerable Mind*. New York: Columbia University Press, 1954.

International Food Policy Research Institute. *Food Needs of Developing Countries: Projections of Production to 1990*. Washington, D.C.: IFPRI, 1977.

Isenman, Paul J., and Singer, H.W. "Food Aid: Disincentive Effects and Their Policy Implications." *Economic Development and Cultural Change* 25(1977):205-37.

Kahn, Herman; Brown, William; and Martel, Leon. *The Next 200 Years*. New York: William Morrow, 1976.

Ladd, Wilbur. "Continental Habitat Status and Long-Range Trends." In *Third World International Waterfowl Symposium* (1978) 14-19.

Laird, Roy D. "Grain as a Foreign Policy Tool in Dealing with the Soviets: A Contingency Plan." *Policy Studies Journal* 6(1978):533-7.

Lee, John E. Jr. "A Framework for Food and Agricultural Policy in the 1980's." Paper presented at the Southern Agricultural Economists Meeting, Hot Springs, Arkansas, 1980.

Levine, Herbert, S.; Rushing, Francis W.; and Movit, Charles H. "The Potential for U.S. Economic Leverage on the USSR. *Comparative Strategy* 1(1979):371-404.

McGovern, George. "Human Rights and World Hunger." *National Forum* 69(1979):8.

Morgan, Dan. "The Politics of Grain," *Atlantic Monthly* 246(1980):29-34.

National Center for Productivity and the Quality of Working Life. *Productivity in the Changing World of the 1980s: Final Report*. Washington, D.C.: GPO, 1978.

Newsweek. 2 June 1980.

O'Hagan, J.P. "National Self-Sufficiency in Food. *Food Policy* 1(1976):355.

Paarlberg, Don. *Farm and Food Policy: Issues of the 1980s*. Lincoln: University of Nebraska Press, 1980.

Paarlberg, Robert L. "The Failure of Food Power." *Policy Studies Journal* 6(1978):537-42.

Pavitt, K.L.R. "Malthus and Other Economists: Some Doomsdays Revisited." In *Models of Doom*. Edited by H.S.D. Cold, *et al.*, pp. 137-58. New York: Universal Books, 1973.

Quance, Leroy. "The Long-Range Outlook for U.S. Agriculture." Paper presented at the annual meeting of the Association of Agricultural Bankers, Alexandria, Va., 22 June 1976. Mimeographed.

Sampson, R. Neil. "The Ethical Dimension of Farmland Protection." In *Farmland, Food and the Future*. Edited by Max Schnepf, pp. 89-98. Ankeny, IA.: Soil Conservation Society of America, 1979.

Schertz, Lyle P. *et al.* "Another Revolution in U.S. Farming?" U.S. Department of Agriculture, Economics, Statistics, and Cooperatives Service. Agricultural Economic Report No. 441. Washington, D.C.: USDA ESCS, 1979.

Schneider, William. *Food, Foreign Policy and Raw Materials Cartels*. New York: Crane, Russak, 1976.

Seevers, Gary L. "An Evaluation of the Disincentive Effect Caused by P.L. 480 Shipments." *American Journal of Agricultural Economics*, pp. 630-42, August 1968.

Sugarman, Robert J. "Institutional Constraints in Managing Resources in North America." In *Resource Constrained Economies: The North American Dilemma*, pp. 49-62. Ankeny, Iowa: Soil Conservation Society, 1980.

U.S. Bureau of Labor Statistics. 1977 Employment Requirements Table for Input-Output Matrix. 1980. Unpublished.

U.S. Congress, House of Representatives. Committee on International Relations. *Use of U.S. Food Resources for Diplomatic Purposes: An Examination of the Issues*, Washington, D.C.: GPO, 1977.

U.S. Department of Agriculture. *Yearbook of Agriculture*, Washington, D.C.: USDA, 1921, 1923, 1958, 1975.

U.S. Department of Agriculture, Economics, Statistics, and Cooperatives Service. *Alternative Futures for World Food in 1985*. Foreign Agricultural Economic Report No. 146. Washington, D.C.: USDA, 1978.

U.S. Department of Agriculture, Economics, Statistics, and Cooperatives Service. *Foreign Ownership of U.S. Agricultural Land*. Second Report to Congress under the Agricultural Foreign Investment Disclosure Act. Washington, D.C.: USDA, 1980.

U.S. Department of Agriculture, Soil Conservation Service. *National Resource Inventories 1977, Final Estimates*. Washington, D.C.: USDA, 1979.

U.S. Department of Agriculture, Soil Conservation Service. *RCA Appraisal. Review draft, part 1*. Washington, D.C.: USDA SCS, 1980.

Wallensteen, Peter. "Scarce Goods as Political Weapons: The Case of Food." In *The Political Economy of Food*, by Vilho Harle, pp. 47-93. Westmead, England: Saxon House, 1978.

Wittwer, S.H. "Future Trends in Agriculture Technology and Management." In *Long-Range Environmental Lookout*, pp. 64-107. Proceedings of a Workshop, 14-16 November 1979. Washington, D.C.: National Academy of Sciences, 1980.

CHAPTER 1
Competition for Land and the Future of American Agriculture

PHILIP M. RAUP

The nature and intensity of competition for land in the United States have changed dramatically over the last three decades. Throughout the era of new land settlement, competition over land use first centered on trees versus crops. Later, as settlement moved west into the Great Plains, it centered on grass versus crops. This era ended in the 1930s, except in timbered portions of the lower Mississippi Valley and scattered areas of the Mountain States and the Northwest. The Taylor Grazing Act of 1934, which mandated that some grassland areas should permanently remain grass, symbolized a policy change in resource management. This coincided almost exactly with the interwar peak in cropland acreage of 384 million acres in 1931-1932 (U.S. Department of Agriculture [USDA], 1978, p. 19).

Although distorted by depression and wars, competition for land from the mid-1930s to the mid-1950s was confined primarily to competition among various sown crops. With forest-farm and cropland-rangeland boundaries reasonably well defined and stable, the land-use arena in which competition occurred was dominated by crop agriculture. Given irrigation and mechanical pickers, cotton could be planted westward of the Old South, in the Texas high plains, Arizona, and California. Because of quick-maturing hybrids, corn boundaries potentially could be moved several hundred miles northward, while the soybean began to emerge as a major competing crop in the Corn Belt.

The major causes of current interest in competition for land had not yet commanded public attention by the end of the Korean War.

The trends and events that generate this interest have multiple roots. The maturity of the United States as an urban society is undoubtedly the most important root. In 1950, only 5 states had over 80 percent of their population classified as urban, and 20 states had half or more of their population classified as rural. Twenty years later, in 1970, there were 12 states with 80 percent or more of their population classed as urban and only 6 states with half or more of their population classified as rural (U.S. Bureau of the Census, 1965, 1979b). In 1960, Standard Metropolitan Statistical Areas (SMSAs) included 8.7 percent of the land area of the United States. By 1974, this percentage had nearly doubled, with SMSAs accounting for 16.7 percent of the total area (Coughlin, p. 30). In 1980, 1/5 of the land area of the contiguous states is estimated to be within the boundaries of SMSAs (*New York Times*, March 24, 1980).

The rapid expansion of urbanizing areas was associated with an accelerated conversion of agricultural land to nonagricultural uses. Using a stratified sample of nine locations per county (typically quarter-sections of 160 acres) for 506 counties, the Soil Conservation Service estimated that 16.6 million acres had been converted to urban uses between 1967 and 1975. An additional 6.7 million acres of land had been converted to water (USDA, 1977b, p. 16). This total of 23.3 million acres converted to urban uses or water in eight years seems to be the source of a frequently quoted estimate that "each year three million acres of farm land are lost to development" (*New York Times*, June 18, 1979a). It is important to note that 4.8 million acres or 29 percent of the 16.6 million acres converted to urban uses were classified as cropland. This yields the less startling but still significant estimate of 600,000 acres of cropland transferred to urban uses annually between 1967 and 1975 (USDA, 1977b, p. 1). Although this is only 0.15 of 1 percent of the 400.4 million acres classified as cropland in 1975, the highly aggregated nature of the data masks the impact of this steady loss of cropland on specific regions and localities. It is the irreversibility of the conversion, and not its magnitude alone, that provides the strongest root for current public concern over the nature of competition for land.

In acre terms, a less visible but much more significant shift in competition for land has been generated by the rapid growth of agricultural exports. In 1950, crops grown on 50 million acres were exported, equivalent to 14.5 percent of the cropland harvested.

Export crop acres doubled by 1975 to 100 million and by 1978 reached 133 million acres or 33.6 percent of the 336 million acres of harvested crops in that year (USDA, 1980a, p. 18). In the marketing year 1979/80, wheat exports are projected to equal 62 percent of 1979 wheat production. Comparable figures for corn are 31 percent; for sorghum, 34 percent; and for barley, 13 percent (USDA, 1980b, p. 21). In fiscal year 1979, soybean exports were 56 percent (USDA, 1979a). In aggregate terms, one of every three crop acres produces for export. For wheat, soybeans, cotton, and rice, the proportion is one acre out of every two, or higher. In terms of competition for land, we have reached a degree of agricultural export dependency for which parallels can be found only in the Antebellum cotton South, or in our Colonial era.

The concern raised by this degree of exposure to world markets is the second major factor supporting current interests in the changing nature of competition for land. Export demand has brought reserves of cropland into production on a scale that has largely eliminated any cushion or margin of safety that might otherwise meliorate fears generated by conversion of cropland to urban and nonfarm uses. Our export successes intensify urban conversion fears. They also have contributed heavily to increases in farmland prices to levels that threaten to prevent an orderly succession in ownership and control of land resources. The twin components of urban and export demand for land induce fears that focus not only on the acres thus preempted, but also on the stability of the structure and organization of agriculture.

These elements in the pattern of competition for land are not new, but they have reached new levels of intensity. To them must be added elements that are new and that derive generally from the interest in biological solutions to energy problems. The potentials for conversion of corn into alcohol, manure into methane, and biomass into energy have captured the imagination of both farm and nonfarm people. To farmers, this potential offers the prospect of demand expansion on a scale that evokes images of a "green OPEC." To conservationists, it seems to offer a realistic substitution of renewable for exhaustible energy sources. Whatever the outcome of current efforts to make the technology of crop and residue conversion economically feasible, it is clear that any successes will involve large acreages of land. This adds an intriguing but largely incommensurable element to concern over competition for land.

A NATIONAL OVERVIEW OF
LAND-USE CATEGORIES

In areal terms and approximate magnitudes, 1/3 of the 2,264,000,000 acres of land in the United States is forest land (32 percent), 1/4 is pasture and range (26 percent), 1/5 is cropland (21 percent), 1/12 is devoted to "special uses"—including urban, transport, recreation, wildlife, farmsteads, and various public installations—(8 percent), and the remainder (13 percent) is in marshes, swamps, rocky or desert areas, tundra, or other lands of low agricultural potential.*

Although land is the most fixed of resources and the land-use categories are broad, it is surprisingly difficult to construct an accurate time-series of land-use statistics. Land-use classifications are cultural as well as economic variables. New uses arise (wilder-ness, wildlife, and recreation areas); old uses acquire new meaning (rivers or lakes become reservoirs or flowage areas); the boundaries separating land uses become blurred.

For the limited purposes of this paper, initial attention will center on agricultural lands and on lands in "special uses"—the urban complex that includes recreational, transportation, rural residential, and institutional uses. For ease of exposition, these two uses can be classified as agricultural and urban uses.

Estimates of the area in agricultural use are reasonably comparable for the period since 1910, especially for cropland. Estimates of pasture and range land exhibit greater variability, due primarily to confusion over the classification of grazing lands in forest areas. Recognizing these limitations, it is instructive to note that the acreage of cropland used for crops in 1978 (368 million acres) was identical with the acreage in 1920-21. In the interim, the acreage used for crops had reached a high of 384 million acres in 1931-32, dropped to 363 million acres in 1939, climbed back to an all-time high of 387 million acres in 1949, and held steady around 380 million acres in 1955, only to decline steadily to a low of 331 million acres in 1962 (almost identical with the 330 million acres of 1910). In Figure 1, this series is plotted since 1910. Measured in national aggregates, the acreage of cropland used for crops has been relatively stable for almost 70 years (USDA, 1980a, p. 19).

The stability is misleading, however. National data mask regional shifts of critical magnitude. Two-dimensional data couched in acres

---

*These data and those that follow on land use by regions and states are drawn primarily from H. Thomas Frey, *Major Uses of Land in the United States: 1974*, USDA, ESCS, Agricultural Economic Report No. 440, November 1979.

leave unreported the enormous changes that have taken place in land-use intensity. Consider first the regional shifts from 1939 to 1978.

CROPLAND SHIFTS BY REGIONS

Approximately 1/3 of the cropland has disappeared in two regions, the Northeast (down 32 percent) and the Southeast (down 35 percent). Losses since 1939 have been only slightly less severe in Appalachia (down 22 percent) and the Southern Plains (down 28 percent).

All other regions gained. The Lake states showed the least change (up 2 percent, 1939 to 1978), followed by the Mississippi Delta states (up 4 percent), the northern Plains (up 12 percent), the Corn Belt (up 17 percent, all since 1970), the Pacific region (up 18 percent), and the Mountain states (up 40 percent). For the Pacific and Mountain regions, virtually all of the increases came in the decade 1939-1948. In those regions, there has been virtually no change since 1950.

The patterns of loss have been similarly varied. The Northeast lost 37 percent of its cropland in a steady decline from 1939 to 1969, followed by a modest recovery in the 1970s. The pattern was the

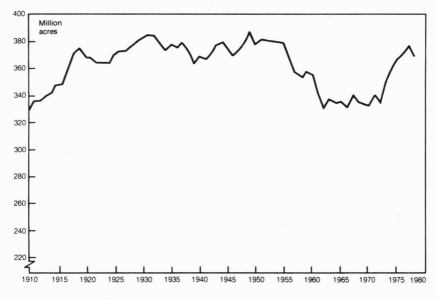

SOURCE: USDA, 1980a

Figure 1. Cropland Used for Crops, United States, 1910–1978

same in Appalachia, where the loss until 1969 was also 37 percent, followed by a bit more vigorous recovery in the 1970s. The most dramatic changes took place in the Delta states region of Mississippi, Arkansas, and Louisiana, which lost approximately 1/3 of its cropland in the 20 years from 1939 to 1958. This region recovered more than all of its loss in the next 20 years, 1959-1978, ending the four decades 4 percent above the 1939 level.

The most acute decline has been in the Southeast. From 1939 to 1969, the region lost 49 percent of its cropland used for crops. The recovery in the 1970s has been significant, but the region still emerged in 1978 as having experienced the largest percentage decline in cropland used for crops of any region since 1939. This may come as a surprise to those accustomed to think of the Northeast as the region most afflicted by loss of cropland in recent decades. The data illustrate an important point: urbanization is not the only reason for cropland declines.

A study by the USDA used aerial photographic interpretation of land-use changes in 53 rapidly urbanizing counties, 1961 to 1970. Over this time period, 35 percent of the land converted to urban use was cropland (Ziemetz, 1976, p. 15). Note that the study period fell within the years of greatest decline in the national total of cropland used for crops, which, for the 48 states, fell from 387 million acres in 1949 to 332 million acres in 1970. There are no recent and comprehensive data to measure the extent to which urban conversion of cropland has been affected by the increased demand for cropland following 1972. Our cropland conversion statistics are primarily "pre-OPEC" and "pre-Russian-grain-sale" data.

The Northeast contains 5 percent of the land area of the United States, 3.6 percent of the cropland used for crops, and 20 percent of the national urban area (Frey, 1979, pp. 18, 26). For this reason, the competition between urban and agricultural land uses in its most concentrated form is centered in this region. For individual states, the loss has been acute. In the national context, the loss of 6 million acres in the Northeast, 1939 to 1978, is 1.6 percent of the 1978 area of cropland used for crops. In contrast, the cropland expansion in 10 years in the Corn Belt, 1969 to 1978, was 12.5 million acres, or more than twice the loss over 40 years in the Northeast. This teaches an important lesson. The trends that we have noted can only be assessed in the context of interregional comparisons.

Cropland losses are not simply a consequence of competition between urban-industrial uses and agricultural uses. Land-use conversion is complex and can be misinterpreted if viewed as a process in which farms are transformed into housing estates or shopping centers. One of the clearest illustrations of this complexity involves forest land.

From 1952 to 1970 the area of commercial timberland (defined as capable of an annual growth of at least 20 cubic feet per acre) declined in all regions of the United States, except in the New England, Middle Atlantic, and South Atlantic states, and in Ohio and Kentucky. The largest increases were in Pennsylvania (2,904,000 acres) and New York (2,537,000 acres), with increases of approximately 1,000,000 acres or more in West Virginia, Georgia, Alabama, and Ohio (USDA, 1973, pp. 227-30).

Approximately 3/4 of the total area of commercial timberland in the United States is now in the eastern half of the country. In 1970, commercial timberlands covered "over 80 percent of the total land in New England and more than half of the area along the Atlantic coast" (USDA, 1977a, p. 15). It isn't a spurious correlation to note that the two regions that experienced the largest percentage losses in cropland since 1939, the Northeast and the Southeast, are also the two regions with the highest proportion of their total land area in commercial timberland.

While we lack data on the long-run historical trend of land-use shifts on specific tracts in these regions, it is clear that, in aggregate terms, cropland losses have been timberland gains. Cropland declines in New England began 150 years ago—the Boston Hinterland was at its agricultural land-use peak in the 1820s. This process was repeated a century later in the Southeast. It is an irony of history that some of the same industries (e.g., textiles, furniture) that drew New England labor off the farms after the opening of the Erie Canal in 1825 were the ones that migrated to the Southeast and repeated the process after the development of the Tennessee Valley Authority (TVA) in the 1930s. In both areas the major causal factors were cheap energy and local supplies of redundant agricultural labor.

The land-use lessons from American history have been unambiguous. Industry has been the chief competitor for farm land, measured not in acres used but in labor force withdrawn. The associated urbanization has generated demands for open space, for recreation, and for residential land uses that are inextricably combined with the largely unplanned expansion of private, noncommercial forest land. Our most urban and industrial regions have become the most heavily forested. In the competition for cropland, the message to date is clear: local trees have been preferred over local food.

In tracing long-run regional trends in competition among major classes of land use, the greatest uncertainty relates to range and pasture lands. Because some cropland is frequently used in rotation with pasture, and some forest land is regularly grazed, it is not even possible to derive accurate figures for the area of pasture and range land. No single government agency is responsible, and no comprehensive national inventory of pasture and range land has ever been attempted.

Measured in acres, the magnitudes of pasture and rangeland are substantial. "Cropland pasture" and "grassland pasture and range" in 1974 accounted for 681 million acres or 30 percent of the total land area. If we add to this the 179 million acres of "forest land grazed," the total is 860 million acres or 38 percent of the land area. This is almost equal to the 900 million acres of "cropland used for crops" plus "forest land not grazed" (Frey, 1979, pp. 3-4).

While much rangeland is of low value in agricultural use, pasturing plays a critical role in the nation's meat supply. The dollar value of livestock gains from grazing was estimated in 1970 as almost equal to the total farm value of the 1970 wheat crop (USDA, 1977b, p. 42).

In regional terms, the most important shifts in pasture and rangeland use have occurred in the Southeast. Florida has emerged as a significant ranching state, and "southern" stocker and feeder cattle have become an important source of supply for the large commercial feedlots centered in western Kansas and the southern Great Plains. The reason for much of the large cropland loss in the Southeast, noted earlier, is the shift of former cropland into pasture and forest land uses.

A FUNCTIONAL SURVEY OF COMPETITION FOR LAND

The period since World War II has witnessed the expansion of nonagricultural demands for land on a scale that may prove in the long view to have been episodic. The major factors competing with agriculture for land use are reservoirs, urbanization, recreation, and energy.

Land for Highways

In terms of both direct and cumulative effects, the dominant episode triggering demand for highway land was the Federal Interstate and Defense Highway Act of 1956. This act injected into the land market a nation-wide demand that was immediately important in terms of right-of-way acquisition but much more significant in terms of the restructuring it generated in land uses and land values. To build some 43,000 miles of interstate highway, aproximately 1.8 million acres were directly acquired as right-of-way, and uncounted millions of acres were given access-values that can be likened to a near-instantaneous conversion from agricultural-use value to urban-use value. One of the key features of this new demand element was the speed with which it was introduced.

From 1956 to 1968, there was an increase of 36,000 miles in the nation's primary road system, most of it in interstate highways. The effect of other primary and feeder-road construction was even more concentrated in time. State highway departments had been building an average of about 55,000 miles of road annually during 1950-55. This dropped to about 45,000 miles per year in 1960-64, while highway authorities were preoccupied with the initial construction of the interstate system. It shot up to 80,000 miles per year in 1966 and 1967, remained above 75,000 miles per year through 1970, and held between 65,000 and 70,000 miles annually through 1974 (U.S. Department of Transportation—DOT, 1975, p. 204). Figure 2 illustrates the dramatic impact of this highway construction boom after 1965.

Between 1956 and 1975, a highway construction effort was undertaken that touched every corner of the country, created and destroyed values on an unprecedented scale, and achieved a transformation with economic and social dimensions that dwarf the railroad-building era in the nineteenth century. The construction period of the railway age was spread over 3/4 of a century. Construction in the interstate highway age was compressed into 20 years. Nothing like it had ever happened, and it is not likely to be repeated.

The decline in new highway construction since 1975 has been almost as precipitous as was the increase after 1965. The drop-off in right-of-way acquisition has been so recent that it is not yet reflected in aggregate national statistics. The records reported by

highway authorities, both state and federal, are focused on miles of
new construction or on the number of tracts or parcels acquired
and not on the acreage of land involved. The decline in highway
construction is indicated in Table 1, showing the number of parcels
acquired for the state highway system in Minnesota, 1957 through

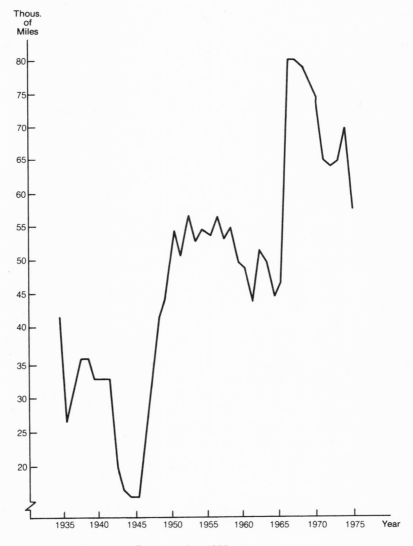

SOURCE: U.S. Department of Transportation, 1975

Figure 2.   Total Mileage Built by State Highway Departments

Table 1.   Highway Right of Way Acquired. Minnesota 1957-79

Fiscal Year	No. of Parcels
1957	2,880
1958	3,375
1959	4,643
1960	2,689
1961	2,631
1962	2,587
1963	1,493
1964	2,219
1965	2,219
1966	2,827
1967	2,516
1968	2,506
1969	2,185
1970	2,219
1971	1,852
1972	1,637
1973	1,567
1974	1,291
1975	960
1976	814
1977	550
1978	785
1979	885

SOURCE: Hansen, 1980

1979. The decline in acres acquired after 1973 is much greater than is indicated by these data. Acquisitions since 1974 have been largely confined to small tracts to complete the system. New construction has been focused on the normal improvement of the regular trunk highways (Hansen, 1980).

There will be a continuing program of new construction to upgrade existing highways. It is quite unlikely that there will be any significant construction of newly located highways in the remaining decades of this century. State highway budgets and the Federal Highway Trust Fund are already overstrained by unexpectedly high maintenance costs. The life expectancy of major segments of the interstate system is proving to be much shorter than originally planned. Given the continuing decline in the number of farms and farmsteads, we may well be on a plateau in the total miles in the nation's system of primary, secondary, and local roads and highways. In any forecast of change, a decline is more probable than an increase.

The competition for land that was fostered by the boom in highway construction is still with us. It will be some years before the echo-effects have been assimilated in land-use patterns. But it seems reasonable to conclude that the major effects are behind us. This is especially pertinent with reference to both the loss of productive farmland and the supporting land-use conversion data derived from the Conservation Needs Inventory of 1967 and the National Resource Inventories of 1977, conducted by the Soil Conservation Service of the USDA. As shown in Figure 2, the period between these two inventories coincides almost exactly with the all-time peak in highway construction activity. Our most comprehensive statistics on farmland losses have been compiled for the period in which the suburbanizing effects of highway programs were at a maximum. This is an unsuitable base for long-range projections.

Highway programs must be given top ranking in any inventory of forces affecting the structure of competition for land in the past quarter century. This will almost surely not be the case in the next two decades.

Land for Reservoirs

In a somewhat longer time frame, dating from the 1930s, a companion record is provided by the loss of agricultural land to dams and reservoirs. The TVA holds partial or full rights to over 1.5 million acres of land, of which an estimated 45 percent or 670,000 acres was prime farmland when acquired (Henderson, 1979, pp. 2-3). This was a significant loss of productive land. But the sense in which the land was "lost" must be qualified. The TVA stabilized the regimes of rivers in the Tennessee Valley, reducing flood damage on adjacent lands. Substitutes for land were created through the stimulus given to fertilizer production and use, with basin-wide planning and cheap electric power.

The situation was reversed in the Missouri River Basin. There, six main-stem dams were formed, creating an almost continuous lake from just west of Sioux City, Iowa, to central Montana. Although the acreage of farm land lost to this chain of reservoirs was substantial, the losses measured in acres alone are seriously underestimated. In this semi-arid farming and ranching country, the traditional farm land-use pattern involved an area of feed-

producing land along or near the river, providing winter feed for
livestock that grazed a much larger acreage stretching for many
miles on both sides of the river. In a representative ranch, 640
acres of river bottom land might supply the guarantee of winter
feed that made possible the summer grazing of 25,000 acres of
low-productivty grassland.

Flooding the lands used for winter feed to create reservoirs
destroyed this land-use pattern along major segments of the Mis-'
souri River. The acres of land under water or in flowage rights-of-
way are an inadequate measure of the extent of this loss. The
reduction in flood damage was important, but the benefits accrued
largely to downstream lands in Nebraska, Iowa, and Missouri. The
costs in land lost to agriculture were paid in the Dakotas and
Montana.

The dams on the Missouri River were designed for flood control,
for hydroelectric power production, and to provide a nine-foot
navigation channel to Sioux City, Iowa. They were not designed to
promote irrigation. In the competition for land, the main-stem
dams on the Missouri River provide a text-book case of disassocia-
tion of costs and benefits, since flood control and navigation were
of principal value to the downstream states.

In the Columbia basin, involving the third great system of dams
and reservoirs constructed since the 1930s, the loss of agricultural
land was less pronounced. Much of the inundated land was of
low agricultural potential, and irrigation was designed into the
system as a major claimant on the newly available water.

Water has been the winner in the competition between land and
water in the TVA, the Missouri basin, and the Columbia basin. In
appraising the effects of this competition on land-use shifts, the
most important point is that our dam-building era, like our
highway-building era, is largely behind us. More dams and more
highways will be built. But there is little likelihood that in the
foreseeable future we will be able to compress into a similar time
frame any comparable programs of long-range capital formation
affecting land use.

Urbanization

The third trend affecting competition for land in the past half-
century is the headlong rush into and out of the cities. There is a
close link here with the unplanned consequences of programs
outlined earlier. The initial Interstate Highway Act of 1956 did not

authorize the expenditure of Highway Trust Fund monies for construction within cities. This was quickly amended, and in practice the major expenditure of funds has been in metropolitan areas. What started out as a highway program to link cities together became a gigantic program that sprawled cities farther into the countryside. A symbolic representation of the land-use consequences is that of a boulder, not a pebble, dropped into a pond.

Information to illustrate this urban concentration of interstate highway expenditures is extraordinarily difficult to assemble. Fragmentary data suggest that at least 2/3 of total expenditures on the interstate system will prove to have been made within urban commuter-belts, and the proportion is quite likely higher. The spread city, the sprawled city, and strip city have become distinguishing characteristics of the American urban mode.

It is this urbanization of the countryside that has evoked the greatest fears about loss of agricultural land. Unlike the direct loss of land involved in highway and dam building, the full effects of this new definition of an urban way of life are still ahead of us.

As a result of this form of competition between the agricultural and nonagricultural uses of rural lands, the distinction between urban and suburban land uses is losing analytical significance. So is the distinction between suburban and sprawled urban patterns of land use. A new form of land use is emerging, which might be called "rural urban," or "diffused urban," or perhaps "agri-urban."

The characteristics of agri-urban land use that are most distinctive are: (1) an intermixture of farm and rural-residential land uses, with no sharply defined boundaries for either use; and (2) a demand by dispersed, residential users for urban-type services that are not needed by farmers and that often impose unnecessary costs on agricultural land users.

The financial base for the provision of services in rural areas historically has been the property tax. The distribution of the property tax burden has been most equitable when the predominant land use has been relatively homogeneous in terms of type and size of land-using units. A mixture of urban and agricultural land uses leads to shifts in the demand for tax-supported services and in the relative burden of costs.

When an intermixture of farm and nonfarm land uses prevails over any extended time period, it places strains on the property tax as a local fiscal support base. As a consequence, the property tax loses credibility, local officials turn to state or federal sources of funds, and the strength of local government is eroded by the loss of an independent financial base. In short, the rural community loses identity.

The images that are called up by this description of agri-urbanization typically involve the Boston-to-Washington corridor, California, and perhaps northern Ohio and Indiana. In terms of the effect on agricultural land use, these are the major areas of agri-urban concentration. In terms of the number of farm people involved and the impact on traditional land-use patterns, the greatest change has been in the South. In 1974, for example, 41.4 percent of all farm operators in the South reported off-farm work of 100 days or more per year. In contrast, only 30.0 percent of farm operators in the North Central region devoted this much time to off-farm work (Carlin and Ghelfi, 1979, p. 271).

Agri-urbanization is also blurring the distinction between "rural" and "agricultural." Many counties from New England through Appalachia and into northern Georgia are rural but not primarily agricultural. Many counties in the industrial Middle West and the Pacific coast region are important agriculturally but are essentially urban in character. As recently as 1940, "farm residents comprised more than half of all rural people. Today . . . farm people make up only 15 percent of the total rural population" (Brown, 1979, p. 284).

The result is a form of competition for land that is no longer measured in terms of acres converted to nonfarm uses. Actual conversion may involve a relatively small fraction of the total area. The relevant measure is the degree of compatibility between farm and nonfarm uses. The effect of most consequence for agriculture is seen in the limitations placed on the size and intensity of farm enterprises. We have a rapidly expanding area in which the types of agricultural activity must conform to nonfarm concepts of appropriate land use. Dust from field cultivation, noise from tractors working at night, odors from livestock, use of toxic chemicals and fertilizers—these are all aspects of modern agriculture that generate resentment or fear in nonfarm rural residents. Above a relatively low density of rural residential land use, these fears become constraints on the farming mode. This restriction is of much greater potential importance than any loss of land in acre terms.

A geographic restructuring of the settlement pattern will be one of the most long-lasting consequences of this rural diffusion of urban competition for land. In keeping with our historical tradition, the determinant force in this restructuring will be the transport mode. From the Civil War to the Second World War, railroads dominated the locational structure of settlements in the United States. Success in the competition for land was determined by farm-to-market access for the products of land. A railroad map was also a map of the location of urban places. To be distant from a railroad was a major disadvantage, and the results were clearly apparent in patterns of land-use intensity and in land values.

The highway era introduced major changes into this structure. Beginning slowly in the 1920s, with a dampened momentum due to depression and war, the full force of this change was not released until after 1945. Interstitial areas within the railroad network were no longer keenly disadvantaged. The automobile and the motor truck created new transport options, with highways providing augmented capillaries for a transport system previously dominated by arteries and veins.

This era lasted only about four decades, from roughly 1915 to 1965. The interstate system of highways has reimposed a greatly reduced structure of arterial transport routes on the settlement pattern, and urbanization is now clustering around these major routes. Although the noteworthy revival of population growth in nonmetropolitan counties after 1970 has many causes, it is one measure of the decentralizing influence of these major highway corridors (Beale, 1976). We can anticipate the emergence of the interstate city.

The urbanization pattern also will be altered profoundly by unique changes in housing demands resulting from the post-war baby boom. From 1945 to the peak in 1957, annual births increased by 56 percent. For 11 years, from 1954 through 1964, there were more than four million births each year. The decline, which began in 1962, was almost equally abrupt, to a low in 1973 that was only 15 percent above the level of 1945. The path traced by this remarkable upsurge is shown in Figure 3.

Figure 3 also shows the same curve moved forward 25 years. With a minor downward adjustment for mortality, over four million young people will reach age 25 in each year of the 1980s. Age 25 is the "nesting age." Age at first marriage in 1979, for example, was estimated at 22.1 years for women and 24.4 years for men (U.S. Bureau of the Census, 1980, p. 1). The demand for housing will experience the same distortions that characterized the demand for schools in the 1950s, and the demand for colleges and other post-high-school educational institutions in the 1960s. This will be followed by a sharp decline in the 1990s. Or will it?

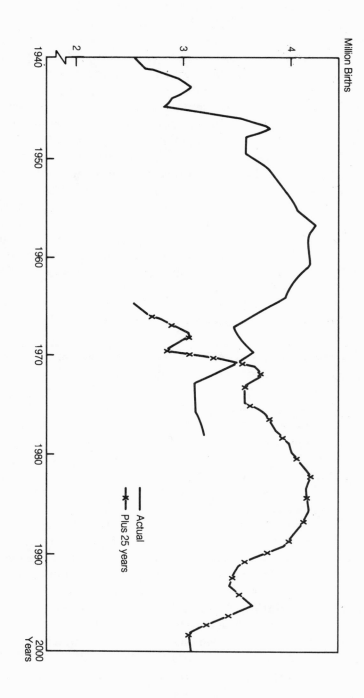

SOURCE: U.S. Department of Health, Education, and Welfare, 1977

Figure 3.   United States Births

What will be the land-use implications of this episodic increase in the demand for shelter? Demand projections based on household formation will almost certainly be wrong. For example, in 1960, "nonfamily" households (those occupied by a person living alone or by two or more persons including no one related to the person maintaining the household) were 15 percent of all households; by 1978 this percentage had risen to 25 percent. Single-person households were 13 percent of the total in 1960 and 22 percent in 1978. Among persons living alone in 1978, 41 percent were over 65 years old, and 79 percent of these were female. As illustrated, the basis for forecasting housing demand has been radically changed (U.S. Bureau of the Census, 1979a, pp. 2-3).

One dimension of the changing structure of the demand for housing has been the sharp drop in the "home-leaving age." This increased the demand for rental housing at a time when inflation-induced advantages of home ownership and the conversion of rental units to condominiums were at a maximum. This "first wave" effect of the baby boom on rental housing demand probably peaked in the latter years of the 1970s. Will it now be superseded by a peak in the 1980s in the demand for single-family detached houses?

If we draw on experience from the 1950s and 1960s, we can project an unprecedented demand for building land. If we project current behavior patterns with respect to marriage and family, we can expect an attenuation of housing demand and a much smaller per capita demand for associated land. This could be associated with a reduction in the frequency of home-leaving at early ages and a continued demand for rental (and hence multiple-unit) housing beyond the nesting ages that dominated housing demand from about 1945 to 1975. This now seems to be the most reasonable projection.

There is some evidence that residential demand for agricultural land is moderating, but this is difficult to document. The major increases in the real costs of credit and energy have occurred so recently that their effects are not yet revealed in comprehensive national statistics.

Data illustrating current trends in the progress of urban sprawl are available for the seven-county area of the Twin City Metropolitan region of Minneapolis and St. Paul. Total building permits peaked at 27,839 in 1972 and fell to 19,774 in 1978. Permits for new construction of single-family residences in rural areas comprised 32 percent of the total in 1973 and 19 percent in 1978. In the first six months of 1979, this percentage had fallen to 15 (Twin City Metropolitan Council, 1980).

We thus may have some basis for adjusting downward the per capita demand for land for housing by a factor of 1/4 or even 1/3. We are still left with the prospect that over 40 million young people will reach age 25 between 1980 and 1990. The resultant demand for residential land in the next 10 years promises to be greater, in terms of area, than any expansion in demand for housing land we have ever experienced in a 10-year period. This demographic phenomenon will be the most destabilizing influence in shaping the competition for urban and rural residential land for the remainder of this century.

Land for Recreation

Although the diffusion of housing demand has been the principal source of professional concern over the loss of agricultural land, it has not generated the political emotions that have been aroused by recreational demand. Voter support for the Vermont Environmental Control Act of 1970 (Act 250), one of the strictest controls on development in the United States, was mobilized primarily by fears of recreational and second-home development and ski resorts (*New York Times*, 1979b). The sharpest land-use controversies in Minnesota have focused on the Boundary Waters Canoe Area Wilderness, contributing to the defeat of a governor and two senators in the 1978 elections. In California, some of the most vehement opposition to land-use control was aroused by establishment of the Whiskeytown-Shasta-Trinity National Recreation Area in 1965, and the scenario was repeated in Idaho with the establishment of the Sawtooth National Recreation Area in 1972. In none of these cases was there a prospective loss of any significant amount of agricultural land. The lesson seems clear: people will express concern over the loss of agricultural land, but they will arise and march when recreational land uses are involved.

The demographic variables outlined above in the discussion of urbanization promise to be of even greater importance in appraising competition for recreation land. Backpacking, canoeing, hiking, and skiing have reached their present levels of popularity coincident with the young-adult phase of the baby boom. These activities are by no means confined to the young, but the strength of the associated demand for land is almost surely correlated with age. We have no historical data to guide us in forecasting future demand. Is it highly elastic, with respect to income, or inelastic? Will people retain their consumption patterns for recreation as family and professional demands on their time increase, or will recreational expenditures be the first to be cut back?

Initial evidence from recent months suggests that the travel-time and distance components in recreational demand for land will be the first focus of efforts to economize. People will seek the same or similar recreational experiences but nearer home. If true, the result could be an increase in recreational demands involving farmland. Parks and recreational areas near population centers will impinge upon agricultural lands with greater force than had been the case when travel was relatively cheap and vacation targets were remote parks or wilderness areas. The most probable consequence of rising energy costs is an intensification of recreational competition for agricultural land.

Land for Energy

Roads, reservoirs, residences, and recreation—these have been overt competitors for productive agricultural land in recent decades. We turn now to a new type of prospective competition that is receiving much attention: land-based production of energy, through the stripmining of coal or the production of alcohols, methane, or related fuels from crops or crop residues.

Strip mining of coal is the most spectacular of these land-based energy forms and has aroused the greatest environmental concern. Total U.S. coal production in 1979 was estimated at 770 million tons, with over 64 percent coming from surface mines. Expansion plans now underway are projected to add 645 million tons of new capacity by 1987, most of it from surface mines, and over half of it in the northern Great Plains (McMartin, 1980, pp. viii-ix). Coal seams in the northern Great Plains range up to 100 feet in thickness, while those in the eastern and middle western states are much thinner—often only two to four feet in thickness. The acreage of land disturbed is in inverse relation to the thickness of the seams. Projections of coal mining activity, 1975 to 1999, estimate 568 thousand acres in average annual use for coal production, of which 358 thousand acres or 63 percent will be in eastern or middle western states.

The return of strip-mined land to agricultural use will be expensive, but in most cases feasible, and the cost is now being incorporated into the price of coal. In approximate terms, the possibility of successful restoration is highest in areas where the disturbance of agricultural uses is of greatest significance. The estimated 658 thousand acres in average annual use, 1975 to 1999, includes an assumption that reclamation of strip-mined land will require 10 years in the Rocky Mountain Region, 8 years in Montana and Wyoming, and 5 years in other strip-mining areas, after the cessation of mining (McMartin, 1980, p. 98).

With these assumptions, and using the average value per acre of gross sales of farm products from surface-mined lands in the various regions in 1974, coal mining from 1974 to 1999 would displace agricultural production worth $16,128,000 annually, at 1974 prices (McMartin, 1980, p. 96). This would have been 0.15 of 1 percent of the 1974 value of agricultural production from lands included within coal-producing regions. It is not possible to defend an argument that coal production poses a threat to food supplies.

Coal mining also threatens water supplies, primarily in the northern Great Plains. To the extent that coal is converted to electric power at the mines, there is also a problem with air pollution and with thermal pollution of water. These problems may prove to be more serious than any impairment to food production capacity through loss of agricultural land.

A much greater potential for reduced food production capacity has been anticipated in the use of grains to produce motor fuel. The current popularity of gasohol in the Corn Belt is understandable in a world of unstable grain prices and unpredictable foreign markets. This popularity also can be traced to the fact that it seems to promise the farmer a recovery of some degree of control over production costs, and a reduction of dependence on suppliers of purchased inputs. It evokes memories of self-sufficiency.

In appraising the potential competition for land from "energy farms" or more generally from the conversion of biomass to liquid fuel, it is essential to keep in mind the true nature of the competition. "Crude oil contains 38 million Btu per ton. Dried plant matter contains 13 million Btu per ton. The energy content falls to about 4 million Btu per ton if biomass is not dried, about the same as oil shale and tar sands" (Zeimetz, 1979, p. 2). Under North American conditions, the competition that gasohol faces is not from other uses of land for crops or biomass but from oil shale, tar sands, or the liquefaction of coal.

Viewed in this perspective, the competition is determined primarily by the relative costs of drying, transporting, and storing alternative forms of biomass. Although large tonnages per acre of wet biomass can be produced in a variety of forms (sugarcane, sweet sorghum, corn, cattails), serious attention has focused on corn. This is because a large fraction of the total energy content of the corn plant is in its grains; corn is relatively dry when harvested; methods of production, harvesting, storage, and transport are well developed; and production is geographically concentrated in the Corn Belt. These are weighty advantages, and they have all but eliminated other crop competitors in the current search for a crop that can provide a liquid fuel to replace gasoline.

But corn still suffers severe disadvantages as a raw material for liquid fuel production in competition with oil shales, tar sands, or coal. These handicaps are of two kinds: temporal and spatial.

The time disadvantage is most easily quantified. Corn is an annual crop and is unsuited to multiple cropping. The entire raw material supply for 12 months of operation of an ethanol plant would be harvested in less than 1 month. The total annual cost of the crop, plus storage costs, would thus involve a capital carrying charge for the feedstock that would largely be absent if oil shales, tar sands, or coal were used as raw material.

In U.S. experience from 1960 to 1973, it cost about 1/6 the cost of the grain to store a bushel for one year (Purdue University, 1979, p. 252). If we take half this amount to cover the storage cost of a supply of feedstock that declines to zero over 12 months, we have a storage cost of approximately 8 percent of the grain cost. If we assume an opportunity cost of capital of 15 percent, then roughly half of this amount would constitute an additional carrying charge over the 12 months. We can conclude that the cost of corn as a feedstock for an ethanol plant would involve additional carrying costs of at least 15 percent of the initial cost of the raw material supply. In many calculations of the cost of producing ethanol from grain, these carrying costs have been ignored.

The spatial disadvantage is also pronounced. Although corn is among the most efficient photosynthetic converters of solar energy, it is distributed quite widely over the landscape. This necessitates substantial transport costs. Using trucking charges prevailing in the Corn Belt in 1980, it is reasonable to conclude that a large-scale ethanol plant using corn as a feedstock would incur transport costs that are 12 to 15 percent of the cost of the corn.

To these estimates of additional costs, which approach 30 percent of the costs of corn production, we must add an insurance factor reflecting the uncertainties of weather and crop yields. Any large-scale ethanol plant would need assurance of a steady supply of feedstock. This would involve reserve storage capacity and a supply territory larger than would be necessary if a constant supply stream could be assumed. Capital carrying charges, storage costs, and transportation costs are thus likely to be larger rather than smaller than the above estimates indicate.

These considerations pinpoint the disadvantages of corn in competition with oil shale, tar sands, and coal. A multimillion-dollar conversion plant could be located at these raw material sites with assurance that a steady supply of feedstock would be available over the depreciable life of the plant (now typically 30 years). The front-end load of capital carrying charges would be known with certainty, as would transport costs. While increases in gasoline costs may make it increasingly feasible to consider grain crops as a source of liquid fuel, they also advance the more likely prospect that attention will turn to oil shale, tar sands, and coal.

There is an additional dimension in the debate over corn as an energy crop that deserves emphasis. Under existing technology, production of ethanol from grain results in a by-product of distiller's grains or "stillage" that has a potential feed value equivalent to roughly 1/3 of the feed value of the grain before distillation. Whether or not energy is gained by making ethanol from grain depends heavily on the effective use of these distiller's grains. This emerges clearly from the conclusions reached in a Purdue University study for the Office of Technology Assessment.

The Purdue study estimated average annual surplus production of grain in the United States over the four years 1976-1979 at 360 million bushels of corn and 260 million bushels of wheat for a total of 620 million bushels. It was assumed that the feeding value of distiller's grains would have been the equivalent of 1/3 of the total grain processed into ethanol. If all of the distiller's grains were used as feed, this would have permitted the processing of 930 million bushels into ethanol without reducing domestic use or foreign exports. This would have produced about 2.5 billion gallons of ethanol annually, equivalent to approximately 2 percent of current gasoline consumption. Assuming a 10 percent ethanol-gasoline blend, this would have supplied 20 percent of the ethanol required nationally.

Using 1978 prices, it was assumed that this grain would have been produced for $2.30 a bushel for corn and $3.00 a bushel for wheat. In order to make the resulting ethanol competitive with gasoline in 1978, the corn would have had to be sold to distillers at $0.70 per bushel and the wheat at $0.75 (Purdue University, 1979, pp. 249-252). Rising gasoline prices would of course narrow this spread, but it also must be assumed that costs of production and processing would rise, although perhaps not as rapidly as gasoline prices.

This example illustrates the potential magnitude of ethanol production from grains without involving a diversion from domestic use or exports. It also emphasizes the highly important role played by the feeding value of the resultant distiller's grains. If these cannot be fully utilized, the net diversion of grain required to produce a given quantity of ethanol will be increased significantly.

In the above example, if the distiller's grains were used at only 50 percent of their feed-value potential, the use of 930 million bushels of grain to produce ethanol would have involved the net withdrawal of 775 million bushels (instead of 620 million), or an increase of 25 percent in the net amount of grain required.

It is improbable that distiller's grains will be utilized as efficiently as would be necessary to justify large-scale production of ethanol from grain. The distiller's grain or stillage emerges from the ethanol plant in highly diluted form, averaging 1,000 pounds of solids in 10,000 pounds of stillage. If it is dried, the energy of drying precludes any net energy gain in the ethanol production process. In wet form, it has a storage life of no more than 24 hours in summer conditions. Because the distiller's grain is largely water, transport costs limit its use to livestock feeders within about 20 miles of the plant. And if fed in large amounts, it constitutes the force-feeding of water, resulting in urinary and nutritional problems in livestock. Its usefulness as a livestock feed is limited to a steady use in the ration, in small quantities (*High Plains Journal*, 1980).

It seems improbable that ethanol from grains can compete with fuels from oil shales, tar sands, or other alternative sources in the foreseeable future, unless it is very heavily subsidized. The most probable outcome is a limited use of gasohol on farms, where the subsidy can take the form of a labor input by the operator, valued at a very low opportunity cost wage, and then only in years of crop surpluses. Gasohol is unlikely to be a serious competitor for cropland.

The same conclusion seems warranted for energy produced from crop residues or farm wastes. Energy from biomass involves the transport or stockpiling of large quantities of low-value raw materials. Manure from large-scale, continuous-flow, confinement livestock-feeding operations is the most promising input. Here, too, successful production seems confined to individual farms and to limited uses. The economics can be compared to a Boy Scout paper drive. If the labor and energy costs of assembling the raw material can be ignored or charged to some other activity, it may be possible to achieve a net energy balance in the actual conversion process. It is extremely unlikely that energy production from biomass will introduce a new element into the competition for agricultural land.

A different situation may prevail with forest lands. Well-developed forests equal or exceed cropland in primary productivity and in annual energy fixation and have the added advantage of a relatively high concentration of biomass per unit area (Leith, 1972, p. 6). They also provide an energy source that is comparatively efficient in direct combustion, thus eliminating the need for processing. The potential of the direct use of wood as fuel is indicated by estimates that currently available aspen within 100 miles of Bemidji, Minnesota, would be sufficient to provide an economic fuel supply for five 25 megawatt electric power plants operating at 80 percent load capacity over a 30-year life (Aube, 1980, p. 27).

The potential for alcohol from forest biomass is also high. If new technologies improve the economic prospects of liquid fuel production from biomass, they seem likely to increase the comparative advantage of trees over annual crops. There is little reason to disagree with Dovring's conclusion that "fuel feedstock from field crops is not likely ever to represent a major contribution to the fuel economy of the United States. Any permanent land surplus available to produce fuel should be planted to trees" (Dovring, 1979, p. 19).

THE CHANGING BALANCE IN INTERREGIONAL COMPETITION FOR LAND

In addition to competition for agricultural lands for nonagricultural uses, there are factors influencing interregional competition. Two factors appear dominant: the demand for irrigated lands and the demand for grain exports. Other factors that are influencing competition are land for reservoir use, urbanization, recreation, and energy.

Irrigation

In terms of cropland acres involved and value of output, one of the most important shifts in land use in the past three decades has been the increase in irrigation. For the United States as a whole, the acreage of irrigated land was relatively constant from 1920 to 1944, at approximately 20 million acres. It jumped 9 million acres by 1954, another 7.5 million acres by 1964, and equaled 41,243,000 acres in 1974, approximately double the 1944 acreage. From 1964 to 1974, over 77 percent of the increase occurred in four states in the central and southern Great Plains: Nebraska, Kansas, Oklahoma, and Texas. Between 1969 and 1974, the increase was confined largely to Kansas and Nebraska. There has been virtually no change in irrigated acreage in the Mountain and Pacific Coast states since 1964; these states still account for over half of the irrigated area of the United States (U.S. Bureau of the Census, 1974 and earlier years).

The output effect of this sharply regional shift in irrigation activity has been confined almost entirely to three crops—corn, sorghum, and alfalfa. These are crops preeminently suited for the production of beef. This is reflected in a massive concentration of beef cattle feeding in large custom feedlots in Nebraska, western Kansas, eastern Colorado, and the panhandles of Texas and Oklahoma. These areas in 1974 accounted for 44 percent of all fed cattle marketed in the United States, while the Corn Belt accounted for only 20 percent (Martin, 1979, p. 100).

The large feedlots are highly concentrated geographically. There are very few north of an east-west line through North Platte, Nebraska, or south of Amarillo, Texas. A circle centered on Garden City, Kansas with a radius of 100 miles encloses approximately 2/3 of the custom feedlots in the region that were actively advertising for business in 1979 (*High Plains Journal,* Aug. 13, 1979).

Many reasons account for this rise of the southern Great Plains and decline of the Corn Belt in cattle feeding. Cattle confined in large open lots are highly susceptible to the climate. Northern lots face severe winters, while southern lots must reckon with heat stress in summer; humid lots greatly increase the possibility of infectious disease. These considerations have been major locational determinants. Other determinants include: the shift of population to the Sun Belt, which has reoriented the market for fed beef; and the conversion of cropland to pasture in the Southeast, and its emergence as the major beef cattle raising region of the nation. There were 9,923,000 beef cows in the Southeast in 1978 and 9,339,000 in the Southwest. These two regions accounted for half of all the beef cows in the United States (Martin 1979, pp. 89-99). Western Kansas feedlots represent a rough approximation of the solution of a gravity model of location for a beef-feeding industry that seeks to minimize the transport costs of its raw material inputs.

These are all important explanations for the restructuring of regional claims upon agricultural output that has resulted from shifts in cattle feeding. In value terms, this restructuring concerns the largest segment of American agriculture. Sales of cattle and calves in 1978 totaled 28 billion dollars or 1/4 of the gross value of agricultural output (USDA, 1980d, p. 9). In acre terms, and with reference to the domestic market, beef feeding is the greatest claimant for the output from harvested cropland. These considerations merit a closer examination of the reasons for the shift.

The major explanation is provided by the development of irrigation in the southern Plains. This provided a rapidly increasing supply of local feed over the past 25 years. But this irrigation has been accomplished at great cost in terms of resources, and the production base is unstable. Virtually all of the irrigation is from groundwater, and all of the groundwater in the area of greatest feedlot concentration is from the Ogallala aquifer. This vast underground lake stretches from the northern Nebraska border to the Texas panhandle, as shown in Figure 4. Its origin is uncertain but apparently geologic. Where there is recharge, it is very slow, especially from Kansas south.

There has been no charge for this water, other than the cost of pumping. As noted earlier, approximately 2/3 of the fuel used in pumping has been natural gas. The water, in effect, has been regarded as a free good, and almost all of the irrigation was developed during a period in which natural gas has been flagrantly underpriced. The water table has been steadily falling; in several Kansas counties it is falling at rates exceeding five feet per year (Kansas Water Resources Board, 1979). Natural gas prices have doubled, and more increases are in prospect. The future of irrigation in the region is entering a critical phase.

A recent USDA study of 32 counties in the Texas High Plains used a simulation model to project irrigation prospects for the period 1976-2025. Applying a conservatively estimated rate of increase for natural gas prices, the study concluded that irrigation

in the region would terminate by approximately 1995. The major
land-use consequence would be a 70-percent decline in grain
sorghum output and a return of wheat as the major dry-land crop
(Young and Coomer, 1980, pp. 27-28).

A contemporary Kansas study attributed 1/4 of the state's
gross farm income in 1977 to irrigation, almost all of which
depends on the Ogallala aquifer (Darling, 1979, p. 90). No estimates of the impact on land use of a decline in irrigation in Kansas

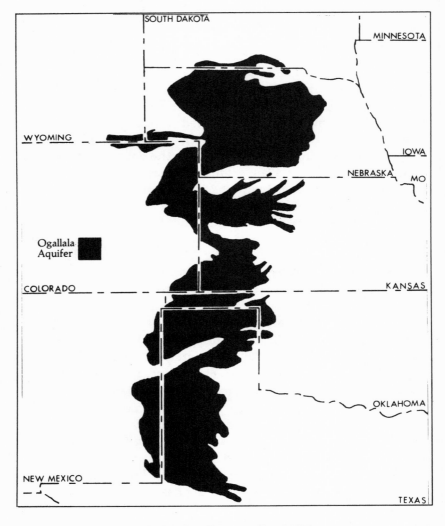

SOURCE: Charles Bradley, Kansas Water Resources Board, 1978

Figure 4.   The Ogallala Aquifer

are available to compare with those from the Texas study, but it must be assumed that the effects will be similar.

These studies raise serious questions about the feed-base for the present concentration of cattle feeding in the southern Great Plains. If rising energy costs and falling water tables make irrigation in the region uneconomic, it will trigger the most significant regional shift in the present geographical pattern of land use that we have in prospect. This would alter the nature of competition for land in the Middle West and Great Plains during the declining phase of Great Plains irrigation as dramatically as it was altered in the expansion phase, and almost as quickly.

The most immediate impact will occur in the panhandle region of Texas, and in eastern New Mexico, where irrigation from groundwater has been increasingly under stress since the 1950s. A reappearance of dryland wheat and unirrigated varieties of sorghum would be associated with a declining feedgrain surplus. If this pattern of land-use change works its way north, as the Ogallala aquifer is gradually exhausted or pumping becomes uneconomic, it will erase the advantage of cheap and plentiful feed that has been the basis for the concentration of cattle feeding in the region.

The feed supply for cattle fed in the southern Great Plains has come predominantly from water withdrawn from an aquifer that is unlikely to be recharged in our lifetime. The one-time withdrawal of this water has permitted the entire increase in feed output to be devoted to beef production, without burdening the feed supplies of the traditional Corn Belt. In approximate terms, the increase in fed beef output of the southern Plains, and this means roughly half of the nation's total supply, has been achieved in the past two decades at no cost in terms of regionally diverted feed grains. The economic rent generated by unpriced water from the Ogallala aquifer and by underpriced natural gas has been capitalized in part into local land values. But in a larger sense, it has been capitalized into a national level of beef consumption that cannot be sustained in the long run without a return to the feed grain supplies of the Corn Belt. We have a fed beef economy that has become dangerously dependent on an exhaustible resource base.

Grain Exports

The most acute competition for land in the United States today is between foreign and domestic producers of meat who use grain as feedstocks. The grains of the Middle West and Great Plains have become the food reserve of the world. Wheat production roughly doubled in 25 years, from an average of 1,077 million bushels, 1951-55, to an average of 2,048 million bushels, 1975-79. Wheat exports in 1978-79 were 1,914 million bushels and were forecast at 1,325 million bushels in 1979-80. Exports have taken more than all of the output increase in the past 20 years.

The record for corn is similar, but the quantities are much larger. On average, the United States produced 2,814 million bushels of corn for grain in 1951-55 and over 7,000 million bushels in both 1978 and 1979. Corn exports in fiscal 1978 were 1,933 million bushels and 2,121 million bushels in fiscal 1979. In dollar terms, wheat exports in fiscal 1979 totaled 4,775 million dollars, corn exports 6,059 million dollars. Exports of feedstuffs (corn, sorghum, and soybeans) in fiscal year 1979 totaled 14,125 million dollars or almost three times the value of wheat exports (USDA, 1980c, p. 21). Feedstuffs have become the dominant agricultural export.

The growth of world demand for feed grains and oil seeds from the United States has generated a massive reorientation of the flow of crops to market. In the nineteenth century, the major export demand for agricultural products from North America was for high-quality wheat, used to produce bread. Since 1945, this trade has shifted to feeding stuffs, and meat has become the final product. This has led to a restructuring of competition for land that is yet to be reflected fully in the structure of American agriculture.

Coincident with this shift of foreign demand from bread to meat, there has been a historic shift in the pattern of transport costs. Over several decades and up to 1979, the real cost per ton-mile of water transport had fallen, and the real costs of land transport had increased. The "continental divide" in rail freight rates for grain in the United States runs through eastern Montana, approximately at the longitude of Forsythe. The cost of grain transport via railroad from there west to Pacific Coast ports is approximately equal to the cost of rail transport east to water transport at Duluth-Superior or at Minneapolis-St. Paul.

In mid-1979, this cost averaged about $1.20 per bushel for wheat. Transport costs by unit trains or river barges from Minneapolis-St. Paul to Gulf Coast ports approximated $0.65 per bushel. The cost of transport from Gulf ports to Rotterdam ranged from $0.25 to $0.45 per bushel. In transporting a bushel of wheat from central Montana to western Europe in 1978-79, approximately 80 percent of the transport cost involved internal transport within the United States.

This transport cost structure has been altered since 1979 by sharply increasing ocean freight rates, reflecting the increasing cost of fuel oil. But it is still true that once a cargo of grain is loaded onto an ocean-going vessel at a U.S. port, it can be shipped anywhere in the world for the approximate cost of transport from central Montana to Minneapolis-St. Paul. The market for North American grain has become a world market, and the revolution in transport costs has contributed heavily to this development.

In appraising land-use consequences of this restructuring of transport costs, an additional institutional phenomenon is important. The most recent innovation in U.S. land-based transport is the "unit train." Shippers can lease entire trains of identical cars, designed to facilitate loading and unloading. The conventional leasing arrangement in 1979 to Gulf ports involved a contract for 45 round trips per year. Shippers who could achieve this minimum of 45 "turn-arounds" in 12 months could obtain substantially lower transport costs.

This introduced a time element as well as a distance element into the market structure for Midwest grains and oilseeds. Transport costs can be reduced if supplies can be located as close as possible to the Gulf, not only because of distance but in order to permit quick turnarounds and thus enable shippers to make the 45 trips per year necessary to qualify for the lower unit-train leasing charges (DeWitt, 1980). This led grain shippers in 1979 to make as many purchases as possible from the lower Mississippi Valley. Corn and soybeans (the principal exports) were procured first from Arkansas, Mississippi, Missouri, Southern Illinois, and Indiana. Procurement then shifted north, in a concentric circle pattern, to include Ohio, northern Illinois, Iowa, Nebraska, and southern Minnesota. Where wheat was involved, Texas, Oklahoma, Illinois, Ohio, and Kansas grain moved to export markets first, followed by grain from Minnesota and the Dakotas, with Montana at the end of the line.

As a consequence, the northern and western Corn and Soybean Belts and the northern Great Plains have become the residual suppliers to the world market for grains and oilseeds. The grain and feedstuffs reserve of the world is stored in this region. Stocks of corn and soybeans in storage in Minnesota on January 1, 1979 and 1980, for example, were at record highs. As of January 1, 1980, 80 percent of the corn and 66 percent of the soybeans were stored on farms (Minnesota Crop and Livestock Reporting Service, 1980). This has created a crisis in farm credit in the region, as producers strive to finance both their grain stocks and the costs of producing a new crop.

It has been conventional in recent years to point out that 1/3 of the crop acres in the United States produce for export. In estimating the participation of each state in the export market, the national percentage of each crop exported has been applied to that state's contribution to total production. This is dangerously misleading in estimating the effect of export demand on competition for land. Any variation in exports will have its greatest impact on producers at the end of the transport line. For example, a variation of $0.50 cents per bushel in the price of corn at Rotterdam in December 1979 would have been approximately 14 percent of the Rotterdam price but 24 percent of the farm-gate price in southern Minnesota (USDA, 1979b, p. 27). Export demand is always of greatest importance to the producers who are most distant from export markets.

These considerations illustrate the regional differences in the effects of the recent expansion of export markets for United States agricultural products. The competition for land in the United States has now entered an international phase. It has been noted that substitutes for land can be found in fertilizers, in superior seedstocks, and in tillage and management practices that obtain more output from the same area.

This situation can be reversed. For foreign buyers, imported grains and feedstuffs are, in effect, a substitute for land, and for the higher levels of intensity that might otherwise be applied to their domestic agricultural resources. In expanding agricultural exports, the United States is "selling" its land at the same time it may be creating substitutes for land through fertilizers and more intensive management practices. In terms of national policy, a key question can be raised: Is this "sale" of more intensive levels of land use through the export of the products of land a wise policy? The superficial response is: Does it pay? That is, do we receive more net benefits through the foreign exchange earned in this way than would be obtained through the use of our land resources for other purposes? If the net effect of an expanded export market for agricultural products is to finance the continued wasteful use of imported petroleum fuels, the answer becomes ambiguous. To date, it is clear that the competition for agricultural land in the United States resulting from expanded agricultural export markets has postponed a confrontation between the true costs and benefits of our current consumption of imported energy. Our agricultural exports, in effect, are financing an increasing portion of an agri-urban life style that depends heavily on the private motor car. Lovers of irony will note that agriculture is thus contributing to the continuation of suburbanizing pressures on rural land, which in turn have generated most of the current concern over loss of farmland to nonfarm uses. Is this the best use of the fertility of the land?

The restructuring of internal locational advantage or disadvantage occasioned by transport cost differentials and the growth of export markets may have an unanticipated consequence for livestock feeding. As noted above, beef cattle feeding shifted from the Corn Belt to the southern Great Plains as a result of rapid increases in the local supply of relatively cheap grains. Hybrid corn, hybrid sorghum, and irrigation facilitated a shift out of livestock in Midwest agriculture that has led one student of the problem to speak of the resulting "grain deserts" (Dovring, 1979, p. 15). The Delta states led in a shift of much of the nation's richest alluvial soils into corn and soybeans. This has resulted in a concentration of very large cash grain farms in the Mississippi Delta as well as in the Great Plains.

When coupled with a freight rate pricing structure that has been altered by unit-train leasing practices, this southward shift of feedstuff production seems likely to focus foreign demand on the lower Mississippi Valley. Cattle feeders in the southern Great Plains will be bidding primarily against foreign buyers for their feedgrains. Relative feedgrain and soybean prices may reflect this shift by making it again attractive for farmers in the Upper Mississippi Valley to feed their grains to livestocks, as was once the ruling case, rather than sell the grain for cash.

There is some evidence that this shift is beginning. The USDA maintains a continuously updated data series on costs of production in the Great Plains and in Corn Belt cattle-feeding enterprises. Throughout 1972 and 1973, the price required to cover all costs of production of fed beef in southern Great Plains feedlots averaged 10 percent below the break-even price in Corn Belt feedlots. Throughout 1979, Great Plains feedlot costs were about 3 percent above Corn Belt costs, and this cost differential seems likely to increase (USDA, 1974, pp. 21-22, and 1980d, pp. 13-14). The export market has now become the most important force affecting interregional competition for agricultural land within the United States.

SOME FUTURE PROSPECTS

In reviewing the past half-century of competition for land in the United States, it is clear that the dominant influence has been generated on the demand side. In spite of a doubling of agricultural output since the Second World War, and major changes in the composition of production inputs, the contribution of land as an input in the production process has remained surprisingly constant. In 1910, labor accounted for 53.4 percent of all agricultural inputs, land 20.2 percent, machinery 8.5 percent, agricultural chemicals 1.7 percent, and all other inputs 16.2 percent. In 1978, labor inputs were 16.0 percent of the total, land 21.6 percent, machinery 31.3 percent, agricultural chemicals 9.6 percent, and all other inputs 21.5 percent (USDA, 1980e, p. 8). Changes on the demand side for land have thus come primarily from outside agriculture or from abroad.

The brief survey attempted in this paper has highlighted some of the growth elements in this expansion of the demand for land. In speculating upon future trends, the data point to two major potential shifters in demand: urbanization and foreign trade. The solution to the housing problem that will be acute in the 1980s will provide the most immediate evidence of the direction that will be taken by domestic nonfarm demand for land. There is some evidence that residential demand for agricultural land is moderating. The major increases in the real costs of credit and energy have occurred so recently that their effects are not yet revealed in comprehensive national statistics.

The short-term prospect is for a substantial reduction in the pressure of urban demand on rural lands. The longer term prospect will be a function of land-use planning and guidance measures that are only now being introduced in many of the most critical areas of urban impact. A forecast of these trends is primarily an exercise of political and not economic judgment.

This leaves the foreign market for agricultural products as the major unknown. To the extent that agricultural land use becomes a tool of foreign policy, we can expect this to be the greatest influence upon competition for land in the United States in our time.

REFERENCES

Anthony, Willis *et al. What to Grow in 1980 in South Central Minnesota.* FM 418.6 SC. St. Paul: University of Minnesota, Department of Agricultural and Applied Economics, 1980.

Aube, Peter J. "Cost-Analysis of a Wood-Fueled Power Plant in Minnesota's Inventory Region 2." Plan B paper for the MS degree. St. Paul: University of Minnesota, College of Forestry, 1980. Mimeographed.

Beale, Calvin L. *The Revival of Population Growth in Nonmetropolitan America.* U.S. Department of Agriculture, Economic Research Service, ERS-605 revised. Washington, D.C.: USDA, 1976.

Brown, David A. "Farm Structure and the Rural Community." In *Structure Issues of American Agriculture.* U.S. Department of Agriculture, Economics, Statistics, and Cooperatives Service. Agricultural Economic Report No. 438, p. 284. Washington, D.C.: USDA, 1979.

Carlin, Thomas A. and Ghelfi, Linda M. "Off-farm Employment and the Farm Sector." In *Structure Issues of American Agriculture.* U.S. Department of Agriculture, Economics, Statistics, and Cooperatives Service. Agricultural Economic Report No. 438, pp. 270-2. Washington, D.C.: USDA, 1979.

Coughlin, Robert E. "Agricultural Land Conversion in the Urban Fringe." In *Farmland, Food and the Future,* pp. 29-47. Ankeny, IA.: Soil Conservation Society, 1979.

Darling, David D. *Earnings Attributable to the Differential Yields from Irrigated Crops.* Bulletin No. 24. Topeka, KS.: Kansas Water Resources Board, 1979.

DeWitt, Lawrence F. Speech at Seminar on *Rural Freight Transportation, New Policy Initiatives,* Agricultural Extension Service. St. Paul: University of Minnesota, January 8, 1980.

Dovring, Folke. *Cropland Reserve for Fuel Production.* Staff Paper 79 E-68. Champaign: University of Illinois, Department of Agricultural Economics, 1979.

Dovring, Folke, and Yanagida, John F. *Monoculture and Productivity: A Study of Private Profit and Social Product on Grain Farms and Livestock Farms in Illinois.* AE-447. Champaign: University of Illinois, Department of Agricultural Economics, 1979.

Frey, H. Thomas. *Major Uses of Land in the United States: 1974.* U.S. Department of Agriculture, Economics, Statistics, and Cooperatives Service. Agricultural Economic Report No. 440. Washington, D.C.: USDA, 1979.

Hansen, A.J. Minnesota Department of Transportation. Personal communication. 28 March 1980.

Henderson, H.A., and Headden, B. "Potential Nonrange Public Lands for Food Production." Paper presented at symposium on *Public Lands Belong to All the People.* Annual meeting, American Association for the Advancement of Science. Houston, Texas, 8 January 1979. Mimeographed.

Highland Plains Journal. Dodge City, KS. 13 August 1979; 1 February 1980.

Kansas Water Resources Board. *Newsletter,* vol. II, no. 2, November 1978; and *Newsletter,* 1979. Topeka, KS.: Kansas Water Resources Board, 1978 and 1979.

Kansas Water Resources Board. *Final Report of the Governor's Water Resources Board.* Topeka, KS.: Kansas Water Resources Board, 1978.

Keefer, Thomas Anthony John. "Forest Biomass Energy Potential." In *Resource-Constrained Economies: The North American Dilemma,* pp. 207-215. Ankeny, IA.: Soil Conservation Society, 1980.

Lieth, H. "Modeling the Primary Productivity of the World." *Nature and Resources.* Vol. VIII, no. 2. April-June 1972.

Martin, J. Rod. "Beef." In *Another Revolution in U.S. Farming?* Lyle P. Schertz *et al.,* U.S. Department of Agriculture. Washington, D.C.: USDA, 1979.

McMartin, Wallace; Whetzel, Virgil; and Myers, Paul R. *People, Agricultural Resour-*

ces and Coal Development. U.S. Department of Agriculture, Economics, Statistics, and Cooperatives Service, Economics Division. Fargo, ND.: USDA, 1980.

Minnesota Crop and Livestock Reporting Service. *Minnesota Crops.* Issue No. 02-80. St. Paul, MN.: Minnesota Crop and Livestock Reporting Service, 1980.

New York Times. 18 June 1979[a]; 22 June 1979[b]; 24 March 1980.

Purdue University, School of Agriculture. *The Potential of Producing Energy from Agriculture.* Final Report to the Energy from Biomass Study, Office of Technology Assessment, U.S. Congress. Washington, D.C.: OTA, 1979.

Twin City Metropolitan Council. "Residential Building Permits by Policy Area." Unpublished data. St. Paul, Minnesota, April 1980.

U.S. Bureau of the Census. "Irrigation and Drainage on Farms." In *Census Of Agriculture: 1974.* Vol. II, part 9. Washington, D.C.: U.S. Bureau of the Census, 1978.

U.S. Bureau of the Census. "Household and Family Characteristics, March 1978." In *Current Popluation Reports.* Series P-20, no. 340. Washington, D.C.: U.S. Bureau of the Census, 1979 [a].

U.S. Bureau of the Census. *Statistical Abstract of the United States.* Washington, D.C.: U.S. Bureau of the Census, 1965, 1979[b].

U.S. Bureau of the Census. "Marital Status and Living Arrangements: March 1979." In *Current Population Reports.* Series P-20, no. 349. Washington, D.C.: U.S. Bureau of the Census, 1980.

U.S. Department of Agriculture. *The Outlook for Timber in the United States.* Forest Resource Report No. 20. Washington, D.C.: USDA, 1973.

U.S. Department of Agriculture. *The Nation's Renewable Resources—An Assessment.* Forest Resource Report No. 21. Washington, D.C.: USDA, 1977[a].

U.S. Department of Agriculture, Soil Conservation Service. *Potential Cropland Study.* Statistical Bulletin No. 578. Washington, D.C.: USDA, 1977 [b].

U.S. Department of Agriculture. *1979 Handbook of Agricultural Charts.* Agricultural Handbook No. 561. Washington, D.C.: USDA, 1979[a].

U.S. Department of Agriculture, Economics, Statistics, and Cooperatives Service. *Changes in Farm Production and Efficiency, 1978.* Statistical Bulletin No. 628. Washington, D.C.: USDA, 1980 [a].

U.S. Department of Agriculture. *Foreign Agriculture Circular, Grains.* FG-24-79 and FG-13-80. Washington, D.C.: USDA, 1979 [b] and 1980 [b].

U.S. Department of Agriculture. *Foreign Agriculture.* Washington, D.C.: USDA, 1980[c].

U.S. Department of Agriculture. *Livestock and Meat Situation.* LMS-195, 1974; and LMS-232, 1980. Washington, D.C.: USDA, 1974 and 1980 [d].

U.S. Department of Agriculture, Economics, Statistics, and Cooperatives Service. *Measurement of U.S. Agricultural Productivity: A Review of Current Statistics and Proposals for Change.* Technical Bulletin No. 1614, Washington, D.C.: USDA, 1980 [e].

U.S. Department of Agriculture, Economics, Statistics, and Cooperatives Service. *State Farm Income Statistics.* Statistical Bulletin No. 627, Supplement. Washington, D.C.: USDA, 1980 [f].

U.S. Department of Health, Education, and Welfare. "Trends in Fertility in the U.S." In *Vital and Health Statistics.* Series 21. Public Health Service. Washington, D.C.: U.S. HEW, 1977.

U.S. Department of Transportation, Federal Highway Administration. *Highway Statistics, Summary to 1975.* Report No. FHNA-HP-HS-S75. Washington, D.C.: U.S. Department of Transportation, 1975.

Young, Kenneth B., and Coomer, Jerry M. *Effects of Natural Gas Price Increases on Texas High Plains Irrigation, 1976-2025.* U.S. Department of Agriculture, Economics, Statistics, and Cooperatives Service. Agricultural Economic Report No. 448. Washington, D.C.: USDA, 1980.

Zeimetz, Kathryn A. *Growing Energy: Land for Biomass Farms*. U.S. Department of Agriculture, Economics, Statistics, and Cooperataives Service. Agricultural Economic Report No. 425. Washington, D.C.: USDA, 1979.

Zeimetz, Kathryn A.; Dillon, Elizabeth; Hardy, Ernest E.; and Otte, Robert C. *Dynamics of Land Use in Fast Growth Areas*. U.S. Department of Agriculture, Economic Research Service. Agricultural Economic Report No. 325. Washington, D.C.: USDA, 1976.

CHAPTER 2

Soil Productivity and the Future of American Agriculture

FREDERICK N. SWADER

The productivity of our soil is one of the most important elements affecting modern American agriculture. Soil productivity, determined largely by the natural characteristics of the soil and by the climate, may be considered a resource and discussed in conventional resource terms. Since soil can sustain agricultural production for many years, soil productivity may be considered a "renewable resource." Conversely, soil productivity also may be considered a "nonrenewable resource" for the following reasons: the amount of soil and its suitability for production is limited, soil is lost through erosion or irreversible conversion, and soil is "renewed" only at geological rates. There is increasing concern that American agriculture has unwisely used the soil and allowed soil erosion and compaction to reduce the stock of soil available for future farming activities. When combined with the fact that an estimated million acres of cropland are converted to other uses annually, there is growing concern about a long-term inability to meet domestic and world food demands at affordable prices.

Some of the natural characteristics of soil can be enhanced: wet soils may be drained; soils in arid regions may be irrigated; soil acidity or alkalinity may be modified as desired; and nutrients may be added in fertilizer materials. These measures can make more cropland available for agricultural production and also can compensate for the loss of natural soil characteristics, which occurs through soil erosion and soil compaction.

As discussed in other chapters of this book, however, these measures to enhance the productivity of soil are limited. For example, the production of synthetic nutrient elements greatly depends on the availability of energy. Irrigation depends on the competition for available water. The development and use of genetic measures will require the continued support of intensive research efforts.

This paper evaluates the productivity of soil in the United States and examines the potential constraints on the maintenance of this productivity. It first discusses the properties of soil that account for its productivity and then examines the problems that soil erosion, soil compaction, and energy production from crop residues may pose for the maintenance of the soil resource. It then discusses the extent to which fertilizers and organic wastes can ameliorate the effects of these problems. The paper concludes with a review of the policy questions associated with soil conservation and some suggestions for government programs.

SOIL PRODUCTIVITY

Soil productivity is the capability of a combination of factors to produce desired levels of yields. Some of the factors affecting soil productivity are:

- The water balance of the soil: its suitability for the species of interest;

- The temperature regime of the soil and its surroundings: the length of the frost-free growing season, the accumulation of heat energy;

- The chemical properties of the soil: its acidity or alkalinity, its content of available nutrients, its content of acceptable levels of potentially toxic elements;

- The physical properties of the soil: its stoniness or the lack of stones, the presence or absence of growth-limiting physical conditions, its depth as it influences its water balance, its texture, and its influence on management requirements.

The interactions among these factors vary widely, creating great swings in global production and subsequent market demand. These factors cannot be ranked in importance, since the factor limiting production may vary from year to year and from place to place. The random nature of external factors—especially rainfall quantity and distribution, particularly in nonirrigated agriculture, and, to a lesser extent, the accumulation of heat energy—makes it important to exercise as much control as is feasible over the internal factors of productivity. Such factors as soil depth, soil tilth, and soil fertility can generally be managed to maintain suitable conditions for productivity.

This paper focuses only on the chemical and physical properties of soil. Since much of U.S. agriculture is based on soils with favorable chemical and physical properties, it is important that these properties be economically conserved and enhanced.

Chemical Properties

The primary chemical properties important to soil productivity are soil organic matter, and the major nutrient elements important to soil productivity are nitrogen, phosphorus, and potassium. Soil organic matter, which is not a nutrient per se, is not supplied by chemical fertilizers and is difficult to replace. The major nutrient elements can be supplied by chemical fertilizers and are not usually difficult to replenish in the soil.

Organic Matter.   Organic matter is the basis of soil productivity. Although the usual level of organic matter in a mineral topsoil is only about 3 to 5 percent by weight, the importance of organic matter is much greater than its low level would suggest. Organic matter provides a source of energy for microorganisms, serves as a granulator of mineral particles, is a major factor in the water-holding and supplying capabilities of soils, and is a major soil source of phosphorus, sulphur, and nitrogen (Brady, 1974).

Organic matter is largely concentrated in the surface soil. This concentration, combined with low density, makes organic matter particularly susceptible to loss by erosion. There is evidence that a relatively greater portion of the light organic matter of the soil is deposited in waterways rather than being deposited on land. This both limits the productivity of the deposited sediment and degrades water quality.

Nutrient Elements.   The three major nutrient elements in soil are nitrogen, phosphorus, and potassium. There are about 25 nutrient elements identified as essential to plant growth. Most of these are required in small amounts and are often supplied in sufficient quantities by the soil. A partial listing of the nutrients in

40 bushels of wheat includes an estimated 50 lbs. of nitrogen, 2 lbs. of phosphorus, 15 lbs. of potassium, 1 lb. of calcium, and 0.03 lbs. of copper.

Nitrogen. Most crops require more nitrogen than either phosphorus or potassium. Crop growth is limited by a lack of sufficient nitrogen more frequently than by any other nutritional factor. Nitrogen is an essential constituent of plant and animal protein. The production of animal-based protein accounts for most of the nitrogen used in U.S. agriculture. The American diet requires an estimated 180 pounds of farm-site nitrogen per person per year.

About 1/2 of the fertilizer nitrogen applied to crops is recovered by crops in the year of application. This relatively low rate of utilization is due primarily to biological transformations, leaching losses, and volatilization (denitrification) (Council on Agricultural Science and Technology [CAST], 1973).

Phosphorus. Although plants require much less phosphorus than nitrogen or potassium, phosphorus plays a vital role in plant nutrition. Crop plants seldom use more than 20 percent of the phosphorus applied as fertilizer in the year of application. This is because phosphorus combines with a number of elements in the soil, forming compounds of low solubility that are relatively immobile. A small portion of the residual phosphorus is available to subsequent crops, but its limited availability makes repeated applications necessary for sustained high yields.

Potassium. Crop requirements for potassium are relatively high, approaching or exceeding the requirement for nitrogen in some crops. Potassium appears to function in a catalytic role in plant growth. Crop plants recover only about 35 percent of the fertilizer potassium applied during the year the crop is grown. Some of the applied potassium is lost through leaching and soil erosion and some becomes chemically unavailable to plants.

The Physical Properties of Soils

The preceding discussion about organic matter and the major nutrient elements presupposes that the soil itself is a relatively important factor in modern agriculture. The soil acts as a rooting medium for crop plants, providing physical support, water, and nutrients via the plant roots. The magnitude of such contributions is governed by both the chemical and the physical properties of the particular soil. The existing technology of managing soil fertility with chemical and organic fertilizers is sufficient to deal with fertility problems in soils that have suitable physical properties.

Soil Classifications. The soils of the United States exhibit a wide range of physical characteristics. Soil types range from the deep loessial (wind-deposited) soils of the Corn Belt, which often have favorable physical characteristics including considerable depth, to the glacial till soils of the Northeast and the North Central region, which often consist of only a few inches of stony, acid topsoil over bedrock or over a dense, crop-limiting soil layer.

The quality of soil resources for agricultural use is commonly expressed in terms of land capability classes and subclasses, which reflect the suitability of soils for most kinds of field crops. Soils are grouped according to their limitations in the production of field crops, the risk of erosion damage they face when they are used for field crops, and the way they are likely to respond to soil conservation treatments.

Capability classes, the broadest groups, are designated by roman numerals I through VIII. The numerals denote progressively greater limitations and narrower choices for practical use. The classes are:

Class I: Soils with few limitations restricting their use.

Class II: Soils with moderate limitations restricting their use.

Class III: Soils with severe limitations restricting their use.

Class IV: Soils with very severe limitations restricting their use.

Class V: Soils that are not likely to erode but that have other limitations, which are impractical to remove, restricting their use.

Class VI: Soils with severe limitations that make them generally unsuitable for cultivation.

Class VII: Soils with very severe limitations that make them unsuitable for cultivation.

Class VIII: Soils and landforms with limitations that nearly preclude their use for commercial crop production.

Capability subclasses, soil groups within one class, are designated by adding a smaller letter "e," "w," "s," or "c" to the class numeral (e.g., IVw). The letter "w" indicates that the water in or on the soil interferes with plant growth or cultivation. Designations "e," "s," and "c" indicate limitations imposed by susceptibility to erosion, by topsoil shallowness or stoniness, or by climate (too cold or too dry), respectively.

For most purposes, it is accepted that croplands in Capability Classes I-III are viable croplands, although with different levels of soil management problems. Croplands in Capability Class IV are considered marginal for cropland, and Capability Classes V through VIII are unsuited to cropland use. Table 1 and Figure 1, which both define the agricultural regions, show the distribution of croplands (Table 1) and the distribution of soils (Figure 1) by regions and by capability classes.

A comparison of the data in Table 1 and Figure 1 indicates that, in every region, there is a concentration of croplands in the better soil areas (classes I-III). In the Mountain region, for example, only 18 percent of the *soils* are in classes I-III, but 75 percent of the *cropland* occurs in those classes. Since these soils are also best suited to nonagricultural uses, competition for their use is likely to be important in the future.

Table 1. Cropland Acreage, 1975, by Region and Capability Classes

(millions of acres)

Farm Production Region	Total Cropland	Classes I-III Acres	%	Class IV Acres	%	Classes V-VIII Acres	%
Northeast	17.3	14	81	2.0	12	1.3	7
Lake States	44.2	39	88	3.7	8	1.5	4
Corn Belt	86.3	80	93	4.7	5	1.6	2
Northern Plains	91.2	78	86	9.0	10	4.2	4
Appalachian	20.7	18	87	1.4	7	1.3	6
Southeast	16.3	13	80	2.7	17	0.6	3
Delta States	20.5	19	93	0.6	3	0.9	4
Southern Plains	40.9	36	88	3.4	8	1.5	4
Mountain	40.1	30	75	7.8	19	2.3	6
Pacific	22.1	17	76	4.3	19	1.0	5
AK, HK, PR, VI	0.7	0.4	57	0.1	14	0.2	28
	440.5	344.4		73.0		16.4	

SOURCE: USDA, 1977

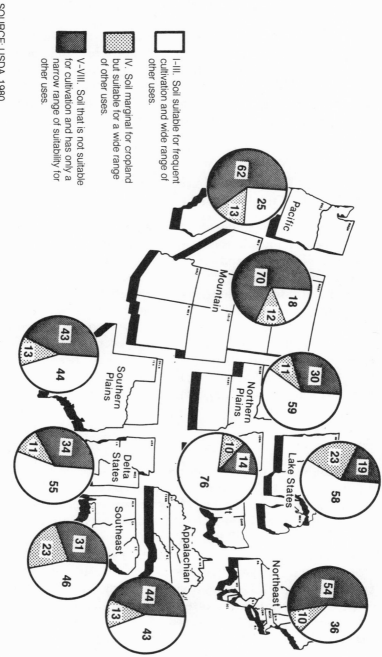

Figure 1. Distribution of Soils by Capability Classes (Percentages) and by Farm Production Regions

SOURCE: USDA, 1980

I–III. Soil suitable for frequent cultivation and wide range of other uses.

IV. Soil marginal for cropland but suitable for a wide range of other uses.

V–VIII. Soil that is not suitable for cultivation and has only a narrow range of suitability for other uses.

When the total acreages of cropland in the various capability classes are compared with the U.S. Department of Agriculture (USDA) projections of cropland needs for domestic and export demand in the year 2000, it appears that there should be an ample supply of cropland to meet these needs. However, these estimates are based on the questionable assumption that the quality of the resource base will remain constant. The rest of this paper discusses the problems associated with the maintenance of productive soil.

CONSTRAINTS ON MAINTENANCE OF SOIL PRODUCTIVITY

Soil productivity is important to the nation's economy internally as it affects domestic food supplies and externally as it affects agricultural exports and the balance of payments. There are a number of constraints on soil productivity, some of which are natural and unmanageable, such as weather conditions; some of which may be lessened, if not completely controlled, such as soil erosion or soil compaction; and some of which are presently more potential than real, such as the use of crop residues to produce energy. The impacts of these constraints have been and continue to be assessed. Understanding the relationships among these various constraints will be important in formulating policies to maintain soil productivity.

Soil Erosion

Soil erosion and its effect on the soil resource base and on agricultural productivity should be a major concern for the future of U.S. agriculture and for the U.S. economy. So that the statistics in this discussion will be meaningful, a short explanation of common measurement units follows.

Soil loss through erosion is usually expressed in the units of "tons per acre per year." A layer of soil covering one acre of land surface six inches deep approximates a soil weight of 1,000 tons. Since a six-inch depth is often the usual depth of plowing, this volume of soil is often referred to as an "acre-furrow-slice." In many soils, this is the most fertile part of the soil. Based on a supply of 1,000 tons of topsoil per acre, the loss by erosion of 167 tons per acre per year would be equivalent to the loss of 1/6 of the

1,000 tons or one inch of soil, removed uniformly from one acre. An annual soil-loss rate of 10 tons per acre is equivalent to the loss of one inch of topsoil every 15 years. Soil lost by erosion may be replaced to some extent by the weathering and amelioration of the usually less productive subsoil, either by natural processes or by cultivation. Natural processes are slow; deliberate cultivation usually requires large inputs of nutrients to increase fertility.

The National Resource Inventories conducted by USDA in 1977 indicate that soil erosion is the major conservation problem on over half of the nation's cropland. Erosion rates exceed 10 tons per acre per year on land in row crops as follows: Southeast, 32 percent; Northeast, 9 percent; and Corn Belt, 19 percent. There are an estimated 283 million acres of cropland with soil erosion rates of 5 tons per acre per year or less; 70 million acres with rates of 5-14 tons per acre per year; and 27 million acres with soil erosion rates in excess of 14 tons per acre per year (Table 2).

Table 2 shows the extent of present sheet and rill erosion on cropland. Soils that are eroding at rates greater than 5 tons per acre per year (97 million acres) constitute about 25 percent of the nation's cropland and about 50 percent of all sheet and rill erosion losses.

One of the concerns associated with the loss of prime farmland to other uses has been that, if export demand continues to grow, the croplands that would be available for expansion are generally more vulnerable to erosion. A study of potential soil losses in the Corn Belt, under a high-export scenario, projected that erosion losses were likely to increase by 50 percent in the eastern Corn Belt states and by 80 percent in the western Corn Belt states. Soil loss increases were projected to range from 40 percent in Illinois to 106 percent in Iowa (Cory and Timmons, 1978).

The rate of soil erosion is estimated by the Universal Soil Loss Equation (USLE), which integrates physical parameters (rainfall energy, soil erodibility, and slope steepness and length) with management practices (contour tillage, terraces, etc.) to estimate the average annual soil loss over several years. Such estimates are compared with an estimated tolerable level of soil loss. The tolerable level (or T value) is the level of erosion that may be sustained without reducing long-term soil productivity. Any erosion rates in excess of the tolerable level (nominally 5 tons per acre per year, but less on shallow soils) are considered excessive and probably represent a deterioration of the soil resource.

It should be noted that the T value assigned to a given soil does not absolutely reflect the rate of physical processes. The T value implicitly includes some of the effects of tillage and other soil manipulation on the transformation of relatively infertile subsoil into a more productive medium. Natural rates of soil weathering are unlikely to produce as much as one inch of new soil material in

Table 2. Extent of Sheet and Rill Erosion on Cropland and the Amount of Sheet and Rill Erosion in Excess of 5 Tons per Acre per Year, by Erosion Interval

Erosion Interval	Thousands of Acres	Cumulative Percentage of Acres	Total Erosion in Excess of 5 Tons per Acre per Year	Cumulative Percentage of Erosion in Excess of 5 Tons per Acre per Year
Tons per acre per year			(1,000 tons)	
0-1	128,186	31.0	0	0.0
1-2	72,596	48.6	0	0.0
2-3	51,619	61.1	0	0.0
3-4	37,060	70.1	0	0.0
4-5	26,693	76.5	0	0.0
5-6	18,661	81.0	7,464	0.7
6-7	13,659	84.3	19,123	2.6
7-8	9,794	86.7	23,506	5.0
8-9	7,667	88.6	26,068	7.5
9-10	6,143	90.0	27,029	10.2
10-11	4,985	91.3	26,919	12.9
11-12	3,754	92.2	24,026	15.2
12-13	3,027	92.9	22,400	17.5
13-14	2,770	93.6	23,268	19.8
14-15	2,403	94.1	22,588	22.0
15-20	7,714	96.0	95,654	31.4
20-25	4,382	97.1	76,247	39.0
25-30	2,891	97.8	64,758	45.4
30-50	5,469	99.1	191,415	64.3
50-75	2,240	99.6	128,800	77.0
75-100	777	99.8	64,103	83.4
100+	712	100.0	168,032	100.0
Total	413,202	—	1,011,398	—

SOURCE: USDA, 1978b

a century. Based on a supply of 1,000 tons of topsoil per acre, if soil erosion occurs at a rate of 5 tons per acre per year, there would be virtually no topsoil remaining after 200 years, in spite of the natural weathering process.

The current estimates of cropland on which soil erosion exceeds the designated tolerance level are shown in Figure 2. Some of the data are questionable because they have been frequently revised. For example, the national average annual rate of soil erosion was estimated at 9 tons per acre in 1978 and revised downward to 4.8 tons per acre only 16 months later (New York Times, May 20, 1979). The significance of such revised estimates is not clear. These estimates probably reflect information about a dynamic situation, gathered as a result of particular program developments. Such data are not collected on a regular basis because of the high costs of collection and changes in the priorities of agencies interested in such matters.

The Impact of Soil Erosion

The effect of soil erosion on crop yields is not easily estimated. Attempts to correlate organic matter losses with soil erosion indicate that the concentration of organic matter in the soil is likely to decrease by about 0.002 percent for each ton of soil lost.

> Losses of organic matter are critical. It is among the first constituents to be lost through erosion yet it is among the hardest to replace. Not only is the soil being depleted of one of its most valuable components, but significant quantities of nutrients, such as nitrogen and phosphorus, are removed with the organic matter. The nutrients can be replaced, but they are far less effective on soils in which organic matter has been lost than they are on soils containing an adequate supply of organic matter. With the trend toward increased fertilizer consumption, additional data are needed relating to specific management practices which will permit a more thorough economic appraisal of organic matter losses caused by water erosion [Barrows and Kilmer, 1963].

The variability in soil properties is important to consider when determining the rates of soil erosion that will not be detrimental to the continued productivity of agricultural soils. Although the deliberate misuse and waste of any resource is deplorable, erosion has fewer direct impacts in soils that are deep, friable, and fertile than in soils that may be shallow, intractable, and infertile. This

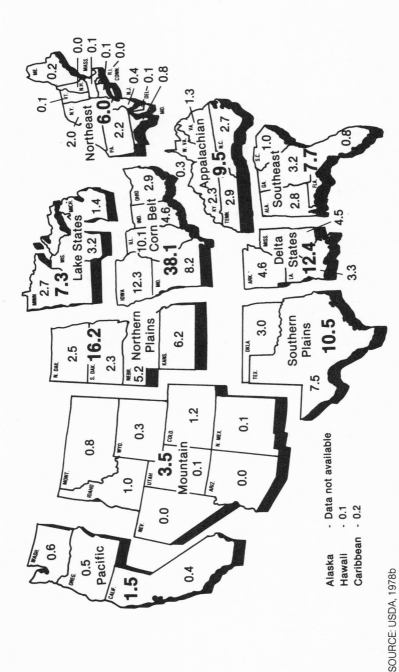

SOURCE: USDA, 1978b

Figure 2. Acreage and Distribution of Cropland on Which the Rate of Sheet and Rill Erosion Exceeds the Soil Loss Tolerance Level, 1977

diversity in soil properties is reflected in the varying effects of erosion on crop yields.

Corn crop yields on land-formed soils of the Atlantic Coastal Plain, for example, varied from 16.3 bushels per acre in areas where 15 cm (6 inches) or more of surface soil had been removed to 135.6 bushels per acre in areas where the original surface soil had been supplemented by additional surface soil. Both the adverse chemical characteristics of the soil environment—low pH, high aluminum saturation, deficiency of plant nutrient and high phosphorus-fixing capacity—and the adverse physical characteristics—crusting and compaction of the shallow subsoil, and plasticity of the deeper subsoil—contributed to the limited crop yields in many altered Coastal Plain soils from which topsoil has been removed (Thomas and Cassel, 1979).

Investigations relating corn production to soil erosion and deposition on soils in the Southern Piedmont region over three widely varying climatic growing seasons found average corn yields of 36, 75, and 92 bushels per acre on soils that had been severely eroded, moderately eroded, or subject to alluvial deposition, respectively. The annual yield reduction calculated at current, nonirrigated corn production levels was 5.8 bushels per acre for each inch of topsoil that was eroded (Langdale, et al., 1979).

Yield reductions due to soil erosion on a Cecil soil in Georgia were determined by comparing crop yields from eroded sites with those from non-eroded sites under similar management systems. For each inch of surface soil lost, yields were reduced by: corn, 2.7 bushels per acre; oats, 3.8 bushels per acre; and seed cotton, 104 lbs. per acre (Adams, 1949).

In contrast to these very regional-specific studies, the USDA (1980) projected the impacts of erosion on crop yields by state (see Table 3). The initial model used for the projections assumed that erosion would continue at the 1977 rates and that technology-associated increases in yields would be 1 percent per year. In view of the 0.2 percent annual yield increases for the 1972-1978 period (Crosson, 1979), the USDA assumption of 1 percent annual yield increases appears unrealistic. The current image of agriculture, the competing demands of other federal programs, and the recent appropriations for agricultural research have led many to predict slower productivity growth.

To illustrate the effects of the disparity between the projected rate and the real rate on the results, assume a corn yield of 100

Table 3. Expected Changes in Yields of Selected Crops in Seven Producing Areas if 1977 Erosion Rates Continue for the Next 50 Years

Crop and Producing Area	Soil Group[a]	1977 Annual Rate of Erosion (tons per acre)	Cumulative Soil Loss over 50 years (inches)	Present Yield (units)[d]	Maximum Potential Yield in 2030[b] (units)[d]	Yield in 2030 if Present Erosion Rate Continues (units)[d]	Percentage of Maximum Yield in 2030[c] (Percent)
Corn 10 (Pennsylvania and New York)	1	2.5	0.8	101	152	152	100
	2	5.5	1.8	81	121	120	99
	3	8.1	2.7	74	111	107	96
	4	9.2	3.1	66	99	91	92
	5	13.6	4.5	67	101	94	93
Cotton 14 (South Carolina and Georgia)	1	3.2	1.1	524	787	787	100
	2	6.2	2.1	388	583	573	98
	3	15.7	5.2	315	472	341	72
	4	22.2	7.4	244	367	256	70
	5	17.4	5.8	227	341	250	73
Soybeans 14 (South Carolina and Georgia)	1	3.2	1.1	24	36	36	100
	2	6.2	2.1	20	30	29	99
	3	15.7	5.2	17	25	21	82
	4	22.2	7.4	14	22	12	56
	5	17.4	5.8	12	18	16	93
Corn 35 (Illinois and Ohio)	1	3.9	1.3	105	157	156	99
	2	4.1	1.4	87	131	128	98
	3	13.2	4.4	76	113	97	86
	4	25.4	8.5	66	99	60	61
	5	42.4	14.1	61	92	50	54
Soybeans 35 (Illinois and Ohio)	1	3.9	1.3	33	50	49	99
	2	4.1	1.4	29	43	43	99
	3	13.2	4.4	24	36	31	87
	4	25.4	8.5	20	31	22	73
	5	42.4	14.1	17	26	17	67
Soybeans 41 (Iowa)	1	3.2	1.0	34	51	51	100
	2	4.9	1.6	29	44	43	98
	3	16.6	5.6	26	39	33	85
	4	18.0	6.0	23	35	21	59
	5	32.2	10.7	20	30	24	79

Crop and Producing Area	Soil Group [a]	1977 Annual Rate of Erosion	Cumulative Soil Loss over 50 years	Present Yield	Maximum Potential Yield in 2030[b]	Yield in 2030 If Present Erosion Rate Continues	Percentage of Maximum Yield in 2030[c]
		(tons per acre)	(inches)	(units)[d]	(units)[d]	(units)[d]	(Percent)
Corn 43 (Illinois and Missouri)	1	4.0	1.3	91	137	137	100
	2	5.1	1.7	74	111	110	99
	3	18.5	6.2	71	107	90	84
	4	14.7	4.9	62	93	76	82
	5	31.5	10.5	50	75	53	71
Cotton 68 (Oklahoma)	1	3.4	1.1	322	483	482	100
	2	3.5	1.2	229	343	342	100
	3	2.9	1.0	218	327	326	100
	4	5.5	1.8	210	315	315	100
	5	2.6	0.9	175	263	260	99
Wheat 68 (Oklahoma)	1	3.4	1.1	28	42	42	99
	2	3.5	1.2	20	31	30	97
	3	2.9	1.0	18	27	26	97
	4	5.5	1.8	17	26	25	96
	5	2.6	0.9	14	22	19	90
Cotton 71 (Texas)	1	4.0	1.3	332	498	497	100
	2	4.3	1.3	247	370	365	99
	3	7.1	2.4	225	338	329	97
	4	6.9	2.3	218	328	322	98
	5	5.0	1.7	206	308	308	100
Grain Sorghum 71 (Texas)	1	4.0	1.3	58	88	88	100
	2	4.3	1.4	44	67	66	99
	3	7.1	2.4	40	60	59	98
	4	6.9	2.3	38	57	57	98
	5	5.0	1.7	38	57	57	100

SOURCE: USDA, 1980

a. Soil groups are made up of aggregations of land capability classes and subclasses in the following manner:

Soil group	Land capability class and subclass
1	I
2	II, IIIs, IIIc, IIIw, IVc, IVs, IVw, V
3	IIIe
4	IVe
5	VI, VII, VIII

b. Based on 1 percent annual increase in yields resulting from technology.

c. Percentages were calculated from unrounded data and therefore may not represent the ratio between the numbers shown for maximum potential yield and eroded yield.

d. Units are in bushels for all crops but cotton, which is shown in pounds.

bushels per acre in 1980. At an annual yield increase of 1 percent, the yield after 50 years would be about 163 bushels per acre. If the annual rate of increase were only 0.2 percent, however, the per-acre yield after 50 years would be about 110 bushels per acre. If corn acreage were to remain at 70 million acres, the disparity in production would be 3.7 billion bushels in the 50th year. Thus, these USDA projections may seriously overestimate the productivity potential in the year 2030.

It should also be noted that the greatest declines in projected productivity occur in soil groups IV and V. The soils in these groups are in capability classes that are generally unsuited to cultivation and that comprise only a small part of the current cropland in most agricultural regions. The extent of existing croplands in these soil groups (see Table 1) ranges from 7 percent (producing areas 35, 41) to 20 percent (producing areas 10, 14). If the current trend toward industrialized agriculture continues, such croplands are likely to be retired from crop production and perhaps replaced by lands better suited to crop production.

Soil Compaction

There is continuing and repeated concern in the popular farm press about soil compaction. The continued expansion of crop acreage to maintain economic farm units has resulted in the use of larger, more powerful farm equipment. This heavier equipment, combined with the pressures of adverse weather at planting or harvest time, creates many opportunities for croplands to become compacted.

This concern is shared by many agricultural researchers, who recognize the trends toward more intensive production of row crops and a concurrent reduction in the use of crop rotations that were once helpful in maintaining good soil structure. Many other causes have been identified, including inadequate drainage and excessive tillage (Robertson and Erikson, 1978).

Compact soil is generally undesirable because plant roots do not function normally in it. To quote Rosenberg (1964):

> Productivity of compacted soils is affected by the increased mechanical impedance, reduced aeration, altered moisture availability and heat flux which follow from increased soil density and reduced pore space. At any one time one or more of these factors may become critical for the growth of plants. Which of these factors actually does become critical will

depend upon the soil type, the climatic conditions, the plant species, and possibly upon the stage of development of the plant when its roots encounter compact soil conditions. Whether a given density increment will hamper or improve plant growth depends then upon whether the soil is looser than, at, or more compact than, the optimal density for the season and stage of growth of the crop growing in the soil. Because of the great complexity of the problem reliable quantitative expressions relating soil compaction to plant growth response are yet to be achieved.

Research data indicate that there is no consistent response of field corn to soil compaction. Canadian data indicate little or no correlation between soil porosity (a measure of soil compaction) and corn yields (Bolton and Aylesworth, 1959). A field compaction experiment in Iowa indicated that soil compaction resulted in reduced corn plant populations and that the compacted areas produced lower yields, even when compared on the basis of equalized plant numbers (Phillips and Kirkham, 1962).

Compaction of two soils (a silty clay loam and silt loam) in Minnesota resulted in slower seed germination, lower plant populations, and delayed maturity. Yields were reduced by 7.5 percent when only surface soil was compacted and by 14 percent when both subsoil and surface soil were compacted (Adams, et al., 1961).

Similar kinds of yield responses have been reported for tomatoes. Root crops and tuberous crops, which form harvestable plant parts directly in the soil, tend to be more consistently sensitive to soil compaction. Potatoes, for example, appear to be sensitive, but this sensitivity varies, depending on the type of potato used in the experiments (Flocker, et al., 1960). Sugar beets show a generally negative response to increasing soil compaction. These decreased yields may be due to excessive soil wetness, insufficient soil aeration, or mechanical impedance (Smith and Cook, 1946).

Cotton has shown varied responses to soil compaction. Yields of cotton have been increased substantially by loosening a dense plow pan or subsoil (Doneen and Henderson, 1953). It is not clear however, whether the major limiting factor was the water supply, the air supply, or mechanical impedance. Research at the National Tillage Machinery Laboratory has shown that cotton yields are often increased by 50 to 60 percent in areas where the soil has not been subjected to tillage-related compaction.* Grain

---

*Dr. W. R. Gill, Director, National Tillage Machinery Laboratory. Personal communication.

yields have been shown to be affected by soil compaction, but the effects were quite variable and dependent on soil texture.

Cropland compaction in California (1971) was estimated to be severe enough to reduce crop yields on 0.8 million hectares. Approximately 0.8 to 2.4 million additional hectares are rapidly approaching the same condition.

As shown above, the effects of compaction are uncertain. Soil compaction clearly affects the transmission of water, gases, and heat; the physical reaction of the soil to applied forces; and the growth of plants. The practical significance of compaction in actual production situations is undefined, however, and its economic impact is unknown.

It has been postulated that the freezing and thawing of soils will mitigate soil compaction. Based on a map of frost penetration depths, and assuming frost penetration deeper than 10 inches to be effective in relieving soil compaction, it was estimated that 30 percent of the nation's croplands is subject to compaction not mitigated by freezing and thawing and that 1/2 of these croplands are used to produce row crops, which normally subject the soil to increased compaction. Thus, mitigation by freezing may occur, but slowly. Ten years of research in Minnesota indicated that freezing was not effective in mitigating the effects of subsoil compaction (Blake, 1976).

The loss in crop value due to soil compaction in the United States was estimated to be $1.18 billion in 1971 (Gill, 1971). In the absence of specific data or more recent projections, it seems reasonable to assume that the annual economic impact of soil compaction exceeds $3 billion (based on the 1971 estimate of $1.2 billion).

Compaction is not easily measured and is even more difficult to recognize. The effects of compaction are not usually dramatic, may not be discernible, and may be very weather dependent. Because most chronic soil compaction is caused by the cumulative effect of accepted methods of tillage and because its avoidance requires innovative approaches (confining all vehicular traffic to permanent roadways), rapid decreases in the extent, severity, and impact of such soil compaction seem unlikely.

Crop Residues, Energy Production, and Soil Productivity

The continuing energy crisis has focused increased attention on the use of organic wastes as fuels and on the conversion of such wastes to fuel materials. Energy from organic wastes has some attractive features. Conversion would reduce our dependence on imported fuels, assist in the utilization of waste materials, provide a "clean" (low-sulfur) fuel, and provide a means of indirectly utilizing the solar energy captured in the wastes.

Agricultural wastes are considered good feedstocks for a waste-to-oil process, with reasonable expectations equaling oil yields of 55 percent and conversions higher than 90 percent. If the estimated annual production of 540 million metric tons of agricultural plant wastes and manure was converted to a fuel oil by a waste-to-oil process, over 220 million metric tons of fuel oil could be produced each year (Steffgen, 1974). Problems with the collection and transportation of these wastes may present major drawbacks to such a conversion, but continued increases in the cost of energy may alter that situation. Federally subsidized programs of biomass conversion also could increase the utilization of such wastes.

In planning waste-to-oil conversion programs, it generally is assumed that all plant residues and animal manures are absolute wastes that have no other uses or benefits, and that constitute net drains on the entire production/consumption system. In fact, this assumption depends largely on the location and concentration of those wastes, as well as on their composition and the alternatives for their utilization. A substantial part of animal manures are reapplied to croplands as a source of nutrients and organic matter. When produced in large quantities in concentrated areas—such as massive feedlots—animal manures may be good sources of raw material for energy production.

Unused plant materials such as corn stalks and wheat straw do not necessarily constitute a waste in the same sense as large accumulations of feedlot manures. These plant residues are often diffused over large areas of cropland. Few crop production systems result in large concentrations of plant materials. The plant residues are often reincorporated into the soil, providing organic matter, recycling many of the plant nutrients taken up during plant growth, and providing some degree of protection against soil erosion by water or wind.

The role of plant residues in reducing soil erosion and the quantities of plant residue that could be removed for energy production were examined for several land resource regions in the Corn Belt and Great Plains (Lindstrom, et al., 1979). Plant residue production in the Corn Belt was estimated at 135 million metric tons per year. This residue consisted of corn residue, 63 percent; soybean residue, 23 percent; and small grain residue, 11 percent. When a constraint was imposed which limited soil erosion to a tolerable level still using conventional tillage methods, only 49 million metric tons or 26 percent of residues were available for energy conversion uses. When various kinds of conversion tillage methods were substituted for conventional tillage, the percentages of available residues increased to levels of 50-60 percent, with 59-78 percent of the cropland area adequately protected. (Additional conservation measures would be required on the remaining cropland areas.) These 49 million metric tons are distributed over some 32 million hectares of cropland.

Plant residue production in the Great Plains amounts to about 7.4 million metric tons per year and is composed of small grains (primarily wheat), 75 percent; corn, 14 percent; and sorghum, 11 percent. When three levels of constraints against soil loss by wind erosion were applied to the model (6.7, 11.2, and 22.4 metric tons per hectare), the percentages of the residues produced that are available for removal are 15 percent, 21 percent, and 38 percent, respectively. If soil erosion losses are limited to 11.2 metric tons per hectare (5 tons per acre), available residues would total about 1.6 million metric tons per year. These crop residues contain plant nutrients that must be replaced if the residues are removed from croplands.

The energy cost of replacing these nutrients must be considered in any energy conversion strategy. Energy is required in producing fertilizer materials as well as in transporting or applying them. The social costs of the increased erosion—and its effects on soil productivity and on environmental quality—also must be considered (Holt, 1979). The resource cost of using natural gas to produce replacement nitrogen or using declining phosphate reserves to replace removed phosphorous also must be considered.

A study conducted by Koelsch, et al. (1977) revealed that, of 154 farmers surveyed in the vicinity of Pratt, Kansas, 60 percent indicated that they were unwilling to allow any of their crop residue removed. The role of crop residues in reducing soil erosion often was cited as a major reason for such unwillingness. The farmers who replied to the survey owned 62,000 acres of wheat. They indicated a willingness to allow straw removal on only 18 percent of the land.

The study also revealed that dryland grain sorghum did not represent a viable source of residues for use as an energy source, that irrigated corn and grain sorghum represented better sources of residues for energy, and that 1/3 of such residues might be removed without greatly increasing soil erosion if slight changes in tillage practices were made. Some additional conclusions of the study include the following.

- In 1977, the cost of wheat straw delivered to the local utility company was not competitive with coal.

- Fields will need to be modeled individually to calculate safe levels of residue removal.

- Development of a minimum tillage program for wheat production is essential for increasing the availability of wheat straw for use as energy.

- An educational program for farmers would be necessary to improve support for a residue removal program. Proven tillage practices that use residue more efficiently in erosion control must be encouraged.

A series of studies conducted by the Science Education Administration-Agricultural Research (SEA-AR), USDA (Lindstrom, 1979) indicate the following.

- About 4 million metric tons of crop residues could be available from 4 states in the Southeast (Campbell, et al., 1979). The density (tons per acre) of such residues was not indicated.

- Three major land resource areas in eastern Oregon could provide 860 thousand tons of available residues annually (in excess of residues needed to control erosion) on 605 thousand cropland acres, for an average of 1.4 tons per acre (Allmaras, et al., 1979).

- Total residues from 9 crops in 20 states amount to 340 million tons that contain 4, 0.6, and 4 million tons of nitrogen, phosphate, and potash, respectively, and that can be recycled. If a major part of these residues is removed for energy conversion, one must consider the consequences of not returning organic matter and nutrients to the soil (Holt, 1979).

The question of "trade-offs" also must be considered. Minimum tillage wheat production techniques have been devised and may offer increased opportunity for residue removal without associated erosion problems. Many of these techniques depend on the increased use of herbicides, which sometimes are perceived to pose some threats to environmental quality.

Should the energy crisis become severe enough, a situation of competing and contradictory federal programs could easily develop, leading to educational programs encouraging the harvest of crop residues at the expense of soil conservation. Such possibilities emphasize the need for a national recognition of the importance of the soil resource and a continuing program for its conservation.

FERTILIZERS AND SOIL PRODUCTIVITY

Fertilizers generally are conceded to be vital to maintaining current and necessary levels of agricultural production. Fertilizers substantially increase yields per acre even when soil productivity has deteriorated due to soil erosion, compaction, or the removal of crop residue. Although questions sometimes are raised about the desirability of a chemical-fertilizer dependent agriculture, there is consensus that it would be virtually impossible for the United States to remain an agricultural exporter without chemical fertilizers.

> Fertilizers are vital to our nation's agriculture. An estimated 30 to 40% of our production is attributable to fertilizer use. The favorable benefit-cost ratio for fertilizer use has resulted in a steady rate of increase in U.S. demand from 8.6 million tons in 1940 to 42.5 million tons in 1973 [CAST, 1973].

The data in Table 4 indicate that, while the harvested cropland acreage in the United States increased by 10 percent from 1965 to

Table 4. Fertilizer Data, Selected Years, 1965-1978

	1965	1970	1975	1976	1977	1978
Harvested Acreage, millions	294.1	283.2	324.0	325.5	332.4	324.8
Fertilizer Nutrients, million of tons	31.8	39.6	42.5	49.2	51.6	47.6
Total Plant Nutrients, millions of tons	10.9	16.1	17.6	20.8	22.1	20.6

SOURCE: Hargett and Berry, 1979

1978, the use of plant nutrients increased by 90 percent during the same period (from 10.9 to 20.6 million tons).

On a regional basis (Table 5), the quantities of total plant nutrients applied have increased, except in New England. The level of nutrients applied results from changes in the extent of harvested cropland within the regions and in the intensity of fertilizer use. The data in Table 6 indicate that there have been increases in the nutrients applied per acre of harvested cropland in all regions. The greatest increases have occurred in the "West North Central" and "Mountain" regions, both of which were applying relatively low rates of nutrients in 1965 (and both of which were substantially below the national average in 1978).

The data in Table 7 provide more specific detail on regional production trends. During the period 1965-1977, the average of harvested cropland declined only in the Northeast, probably as a result of the reversion of some marginal croplands and the conversion of some croplands to other uses. All other regions experienced increases in harvested cropland during the same period, accounting for a 12 percent increase in harvested cropland on a national basis.

The data indicate that there has been substantial growth in the use of fertilizer materials and in farm productivity (outputs/in-

Table 5. Fertilizer Use—Total Plant Nutrients

	(thousands of tons)				
Region	1965	1970	1975	1978	% Change 1965-78
New England	140	126	117	121	− 14
Middle Atlantic	650	731	746	826	+ 27
South Atlantic	1,731	2,081	2,131	2,406	+ 39
East North Central	1,828	3,962	4,344	5,566	+ 97
West North Central	2,233	4,448	5,174	5,908	+165
East South Central	1,015	1,210	1,274	1,550	+ 53
West South Central	1,113	1,779	1,728	1,947	+ 75
Mountain States	359	634	730	806	+125
Pacific	784	975	1,214	1,351	+ 72
United States	10,985	16,068	17,571	20,599	+ 88

Source: Hargett & Berry, 1979

Table 6.   Harvested Acreages[a] and Plant Nutrients[b]

Region	1965	1970	1975	1978	% Change 1965–78
New England					
Acres	2,015	1,439	1,396	1,402	–30
Nutrients	139	175	168	173	+24
Middle Atlantic					
Acres	12,756	11,030	11,788	11,945	– 6
Nutrients	102	132	127	138	+35
South Atlantic					
Acres	15,116	13,873	16,330	16,633	+10
Nutrients	229	300	261	289	+26
East North Central					
Acres	56,498	53,147	61,142	61,136	+ 8
Nutrients	100	149	142	182	+82
West North Central					
Acres	118,419	115,007	131,936	132,217	+12
Nutrients	38	77	78	89	+134
East South Central					
Acres	15,068	14,964	17,993	20,240	+34
Nutrients	135	162	142	153	+12
West South Central					
Acres	38,012	38,151	44,660	42,325	+11
Nutrients	59	93	77	92	+56
Mountain					
Acres	22,976	23,266	25,029	25,092	+ 9.2
Nutrients	31	55	58	64	+106
Pacific					
Acres	13,261	12,183	13,816	13,738	+ 3.6
Nutrients	118	160	176	197	+67
United States					
Acres	294,121	283,180	324,202	324,836	+10
Nutrients	75	113	108	127	+69

SOURCE: Hargett and Berry, 1979

a. Thousands of acres harvested
b. Lbs per acre of harvested cropland

Table 7. Regional Trends—1965-1977

	North-east[b]	Lakes States	Corn Belt	Northern Plains	Appa-lachia	South-east	Delta	Southern Plains	Mountain	Pacific	US
Crop Productivity[a]											
Percent change	0	11	16	36	-12	-15	-21	19	18	31	16
Harvested Cropland											
Percent change	-4	14	18	8	20	28	39	6	4	5	
Fertilizer Nitrogen											
Percent change	54	283	135	268	95	87	62	127	67	131	
Fertilizer Phosphate											
Percent change	10	57	74	137	43	23	63	149	82	68	60
Fertilizer Potassium											
Percent change	26	161	152	508	55	54	84	96	428	137	108
Farm Productivity											
Percent change	8	30	14	20	16	9	20	19	10	29	18

SOURCE: Derived from USDA, 1978a

a. Yields per acre

b. Regional locations shown in Figure 2.

Table 8.   Effects of Elimination of Nitrogen and Phosphate Fertilizer Use on Crop Yields, 1964, Selected Agricultural Subregions

Crop	Subregion	With Fert[a]	Sub-region Average[b]	Without Fert[c]	Effect % [d]
Corn, bu	58 Iowa, Southeast	94.6	88.8	70.8	26.8
Corn, bu	97 California, Central Valley	83.5	82.9	65.0	22.3
Corn, bu	51 Illinois, E. Central	98.2	93.4	59.0	42.0
Wheat, bu	85 Kansas, West & Central	28.5	20.2	18.0	52.0
Wheat, bu	91 Washington, Palouse	48.8	48.8	40.0	18.0
Soybeans, bu	58 Iowa, Southeast	33.6	31.0	31.0	8.4
Cotton, lb	29 Georgia, Coastal Plain	346.0	346.0	212.0	38.7
Alfalfa, tons	89 Arizona, Southern	5.7	5.0	3.3	48.0
Vegetables, tons	10 Delaware	2.1	2.1	1.2	42.9
Vegetables, tons	98 California, Coastal	11.9	11.2	5.8	54.5
Grapefruit, tons	30 Florida, Central	15.7	15.7	1.0	93.6

SOURCE: Viets, 1971
a. Average yield on fertilized cropland
b. Average yield on all cropland
c. First year yield without fertilizer
d. Percentage change in yield for the subregion

puts). The data also indicate relative declines in the growth rates of yields per acre, using 1967 as a basis of comparison. These declines are apparently the influence of weather and a single data-year comparison of yields.*

The persistence and productivity of U.S. agriculture will in part be governed by adequate supplies of fertilizer materials. Net domestic supplies of nitrogen for fertilizer use were expected to total 10.8 million metric tons in 1977-78, including net imports of 600,000 metric tons. Net domestic supplies of phosphate were expected to total 4.8 million metric tons, with a net export of 2.2 million metric tons. Net domestic supplies of potash were expected to be 5.3 million metric tons, including net imports of 3.4 million metric tons (USDA, 1978).

---

*The 1977 crop year was apparently not a good one in Appalachia or the Southeast, which experienced crop productivity index depressions of 11 and 18 points, respectively, below the indexes of 1975 and 1976. In the Delta, 1977 yields improved in 1977 (103) compared to 1975 and 1976 (99 and 96, respectively), but did not compare well with 1965, which was an uncharacteristically high index (125). These aberrations are also emphasized by the increases in farm productivity in the same regions during the same time period.

Fertilizer production requires natural gas or fuel oil; both are used either as chemical constituents or as energy sources. Presently natural gas is the favored form of energy employed in producing nitrogen fertilizers, accounting for about 75 percent of the nitrogen-fixation in the United States. The level of natural gas used to produce nitrogen fertilizers comprises less than 3 percent of the natural gas consumed annually in the United States. Continued shortages and cost increases of both natural gas and fuel oil are likely to be reflected in the price of fertilizer supplies, as are periodic shortages of shipping capacity. Reduced access to abundant fertilizer supplies could greatly affect U.S. production.

Impact of Fertilizers

Estimates of the agricultural production attributable to fertilizer use varied from 20 percent in 1954 (Viets, 1971) to 30 percent to 50 percent in 1973 (CAST, 1973). Viets (1971) modeled the importance of fertilizer to U.S. agricultural production by estimating yield reductions if fertilizer use was eliminated. Table 8 lists examples of the predicted drop in 1965 per-acre yields if all fertilization had been stopped as of 1964. Specific agricultural subregions were selected to represent a wide range of crop production areas and conditions. The estimated effects are based on a one-year discontinuance of fertilizer use.

The data indicate that, after the first year of no fertilization, yield decreases would vary: 22-36 percent for corn; 18-52 percent for wheat; 43-54 percent for vegetables; and 94 percent for grapefruit. In subsequent years, yields would decline further as the residual effects of the fertilizers were dissipated. It should be noted that the "fertilizer effect" was calculated on the basis of the 1964 production practices in the entire subregion, including non-fertilized croplands. The changes due to fertilizer would be more dramatic, in most subregions, if computed on the basis of only fertilized croplands. In addition, it is likely that current effects of fertilizer on crop yields would be even greater, since fertilizer use, in general, has increased since 1964.

The data also illustrate the difficulty of deriving an average figure for the effect of fertilizers on crop production. Instead, this effect must be analyzed on an individual crop basis. There is no effective way to equate bushels of corn to bushels of wheat or to tons of grapefruit except through a common denominator of dollars, calories, or protein (Viets, 1971).

There have only been a few studies that analyze the extent to which fertilizers ameliorate the effect of soil erosion. In one study, the substitution of nitrogen fertilizer for surface soil thickness in Iowa was found to be partially successful, depending on the weather during the particular growing season and the permeability of the soil. The surface soil provided enough nitrogen to produce corn yields of 45-50 bushels per acre more than did the exposed subsoil (Engelstad and Schrader, 1961 a & b). The addition of lime and fertilizer to a Groseclose clay soil in Virginia restored yields of the subsoil to the level of yields in a noneroded soil after four years of treatment (Batchelder and Jones, Jr. 1972). Removal of 30 to 45 cm (12 to 18 inches) of topsoil of a Beadle silty clay loam soil in western Iowa reduced corn yields significantly, although the high application of nitrogen and zinc offset the losses somewhat. The properties of the subsoil made adequate seedbed preparation difficult and resulted in soil crusting, which inhibited corn emergence (Olson, 1977). Deep, calcareous, alluvial Fort Collins soils in Colorado, which had eroded, were found to respond favorably to fertilization. Applications of manure or commercial fertilizers to eroded Fort Collins soils produced practically normal yields of corn, sugar beets, and spring wheat (Whitney, et al., 1950). The differences of these results point out that the ability of fertilizers to ameliorate the effects of erosion is soil-, climate-, and crop-specific.

Improving Fertilizer Efficiency

The continued and increased use of manufactured and mined fertilizers is essential to achieve required levels of crop production. The efficiency of such fertilizer materials may be enhanced by a number of approaches that would also conserve such resources. These include:

- The routine use of soil and plant tissue tests to assess soil nutrient requirements for optimum crop production.

- Improved timing of the application of lime, fertilizer, and other soil amendments to provide optimum fertility with minimum losses due to leaching, surface erosion or evaporation.

- Specialized placement of fertilizer materials to enhance plant uptake and to increase fertilizer efficiency.

- Maintenance of optimum soil reaction (pH) for the particular crop species grown to facilitate crop growth and uptake of major and minor nutrients.

- Improving cropland drainage to allow adequate soil aeration. Excess water in the soil will reduce crop growth, reduce yields, and promote denitrification and subsequent loss of soil nitrogen to the atmosphere.

- Prevention of soil erosion by wind or water to conserve fertilizer nutrients, organic matter, and humus.

- Use of soil and crop management practices that increase crop yields, such as selection of rapid-growth, high-yield cultivars (White-Stevens, 1977).

These measures emphasize that productivity is the biological integration of many factors, including extremely complex technologies.

Alternative Sources: Livestock and Municipal Organic Wastes

The animal population associated with the meat-producing industry generates an estimated two billion tons of raw wastes annually. An estimated 170 million tons of dry manure equivalent is produced in relatively concentrated areas and could be used as a source of nutrients. If all the major nutrients of the manures were retained through the handling process and were available to the fertilized plants, these manures would provide 19 percent of the nitrogen, 17 percent of the phosphate, and 51 percent of the potash applied as fertilizers in an average year.

In 1977, it was concluded that the value of animal manures could not justify much investment in processing, distribution, or hauling costs. Although manures add organic matter and humus to soil and improve its porosity, structure, and its water and nutrient holding capacities, these benefits also can be obtained by incorporating cover crops or crop residues. Manures should, of course, be used as fertilizer materials whenever feasible.

Municipal organic wastes (sewage sludges) also may be used as fertilizer materials, to supplement other nutrient sources, and to recycle the nutrients in the sludges. Concern does exist about the effects of heavy metals in the sludges and their incorporation into the food chain. This seems to be unlikely on most crop lands, since the content of major fertilizer nutrients in sewage sludge is so low as to preclude its practical use on commercial cropland.

A study of nine separate sludge spreading operations found that:

- Sludge management practices were uncontrolled and sometimes inadequate.

- There were no associated human health problems.

- Sludge disposal costs ranged from $8 to $130 per metric ton (dry weight basis) (Otte and LaConde, 1977).

It is estimated that the 1975 U.S. production of major crops —corn, wheat, cotton, soybeans, potato and hay—removed, in a single year, 10 million tons of nitrogen, 3.5 million tons of phosphorus, and 6 million tons of potash. On the basis of these figures, the nutrient content of sewage sludges, and the associated problems, White-Stevens concluded that: "Organic wastes are economically, logistically, nutritionally and practically insufficient and inadequate to meet the present needs of U.S. crop production" (1977). The use of sewage sludge to maintain soil productivity may increase. The likelihood of substantial contributions to soil productivity for the production of food crops, however, appears to be slight.

PROGRAMS, POLICIES, AND PROBLEMS

Soil productivity has been and will continue to be influenced by public programs. Traditional soil conservation programs, which began in the 1930s, were based on limiting soil erosion and its effects on productivity. This approach was developed with the rationale that many soils will not respond rapidly to inputs of fertilizers and organic matter; that such inputs are costly; and that it is more prudent to conserve resources that exist than it is to use other resources to replace those lost by erosion.

These programs will continue to be scrutinized, and questions of program effectiveness will be raised. The answers, as perceived by the public and the Congress, will profoundly affect future USDA programs. Some of the questions likely to recur are the following:

1. What are soil conservation programs? Hoeft (1979) points out that some programs were originally intended to be income redistribution programs for the agricultural populace; that they

were called conservation programs as a matter of political convenience; and that they may continue to exist to serve the first function while named after the second one. Some conservation programs have increased land productivity and current output and have conflicted directly with other USDA programs that are aimed at adjusting current output to market needs at desired prices (Held, 1965).

2. How effective are soil conservation programs? The dimensions of the soil conservation problem and the success of soil conservation programs are being assessed continuously. Two contrary views may be derived from the same data: (1) soil conservation efforts have successfully reduced erosion by one billion tons per year or (2) soil conservation efforts have been only 25 percent effective and have only reduced erosion from four billion tons per year to three billion tons per year (Barlow, 1979). The first approach extols the efficiencies of the existing system; the second approach questions both the system and its efficiency. There is no common agreement on the effectiveness of existing programs or on their impacts on the public as taxpayers or as consumers.

3. How great is the problem? There is no uniformity of terminology or agreement about the magnitude of the problem. Many quotable numbers are used to describe erosion losses; four billion tons per year is currently quite common. Do such figures represent real situations of concern to society or are they an aggregation of casual estimates? The question of how much erosion could be prevented, and at what cost, has not been resolved. It may be much more effective to discuss the problem in terms of the acres of land (or cropland) that are subject to excessive erosion.

4. What are the impacts on crop yields? Although there are data indicating that crop yields decline by some finite amount for each inch (or ton) of topsoil loss, there are few data to convince the public that technology cannot compensate for such erosion losses, or that it is essential to conserve the soil resource. The data are usually complicated by advances in technology, weather conditions, and other variables.

The answers to these—and related—questions are likely to be reflected in U.S. agricultural policy (or the combination of policies that comprise a de facto agricultural policy). Some of the likely policy questions may involve:

1. Value of the soil resource. There is no national policy that places a value on the soil resource or on conserving that resource.

The value of soil is dependent on many variables that affect the value of agricultural products. Heady and Timmons (1975) observed that:

> The potential for U.S. land resources to serve as an adequate base for our own food needs, contributing to world food supplies, and allowing land to be diverted to other priority uses depends upon variables such as population, per capita income, food preferences, agricultural technology, national environmental constraints, and policies that allow a more efficient allocation of land and water resources.

2. Export policy. Can agricultural production continue to finance a national deficit in the balance of payments, and should it be used for that purpose? It appears that there is no immediate alternative and perhaps no immediate need for an alternative. What role can/should agricultural exports play in foreign policy?

3. Environmental concerns. There is no doubt that agriculture has various impacts on water quality. The extent and severity of the impacts have not been quantified. Conservation tillage, which may be the most effective means to reduce erosion, requires the use of herbicides, which may reduce water quality when (if) they reach surface water or groundwater. Although the intent of Congress in P.L. 92-500 ("to restore the biological integrity of the Nation's waters") is clear, there is no clearly defined mandate to protect our nonrenewable soil resources. Coordinated programs should recognize that the nature and severity of the problems will vary from place to place, and strategies to correct such problems must be flexible.

4. Costs of soil conservation. It is not clear who should bear the costs of soil conservation practices, or even what those costs are. The public is not willing to bear all the costs (Libby and Birch, 1979). Some costs will require more time to recover than the farmer's normal planning horizon and should be shared by society. A good case can be made for state or local contributions to reducing soil erosion (Hoeft, 1977) because of the resultant reduction in local expenditures to remedy the damage caused by soil erosion. A policy of consistent federal sharing of such costs would undoubtedly be helpful.

5. Longevity of conservation practices. Agriculture is a dynamic industry. Changes in production methods, in equipment size, in market demand, and in production technology affect the accept-

ability and endurance of conservation practices. As farmers adopt new technology, it may be necessary to redesign and reinstall conservation practices, at some finite cost to both the farmer and society (Sampson, 1978).

6. Conflicting programs and regulations. It is common to find conflicting programs, which reduce the effectiveness of conservation programs. Farmers who have, in the past, planted croplands to sod crops to reduce soil erosion have subsequently found their cropland acreage allotments diminished by the acreage planted to a soil-conserving rotation. Similar situations have emerged concerning the drainage of wet cropland that is converted to meadows or pastures. Large acreages of wet croplands have been included in crop rotations in the past, largely because of USDA programs. Many farmers may be penalized for adopting soil conservation practices.

7. Long-term policies. It may not be feasible to formulate or implement long-range production policies. Wide swings from production to conservation, in response to market demands, may be unavoidable. But such wide swings should be recognized as policy decisions, and both the immediate and long-term costs should be considered.

> In 1975, U.S. farm policy encouraged full production by U.S. agriculture. USDA was not paying farmers to divert productive land from crops, but encouraging farmers to produce what was needed by the world; and had lifted the artificial constraints of acreage allotments and marketing quotas to achieve maximum productive efficiency and lowest possible production costs [Yeutter, 1975].

Some of the costs of such a policy include the sacrifice of some of the soil resource for increased crop sales and perhaps an instantaneous write-off of previous federal investments in conservation practices in the major producing areas. If U.S. farm policy precludes isolating farmers from market demand and encourages them to respond to supply and demand in the world market, the same policy should recognize—and bear—the resource costs, as well as the responsibility for such costs.

8. Transport and energy allocation. Most years, agriculture is hampered by a lack of transportation, either to distribute fertilizer materials from manufacturing centers to farm regions or to transport grain from farm regions to shipping ports. In any future energy crisis, consideration should be given to energy allocations

to agriculture for fertilizer production and distribution as well as for crop production.

9. Preservation of good farmland. Sufficient lands are available to produce both domestic needs and exports well into the future. But the conversion of one million acres of good farmland each year to other uses will ultimately pose problems for producing sufficient quantities for both markets. Policies to preserve farmland should be continued and encouraged at the state and local levels.

CONCLUSIONS

Agricultural production remains the basis of our nation's economy. It may be argued that there is sufficient global agricultural productivity to support the U.S. population even in the absence of a viable national agricultural industry. The realities of supply and demand, price competition for available resources, transportation and distribution inefficiencies, and government policies make it clear that the absence of a national agricultural industry would dramatically and profoundly affect the entire world, and that the greatest impact would be felt in the United States.

The importance of agriculture to the U.S. economy is recognized by those associated with agriculture. It is less clear that the general, voting public has similar perceptions. The lack of public perception may be reflected in reduced budgets for agricultural research. There is serious concern that the recent low rate of crop yield increase may be a product of the reduced level of support for agricultural research in the past decade. Most of the best land available for crop production is now being used for crop production. Future increases in productivity are likely to be relatively expensive, and the likelihood of an efficient expansion of agriculture seems small.

A recent poll (USDA, 1980) indicated that many Americans favor programs to conserve the soil resource. The soil resource is only one (admittedly vital) component of the factors of soil productivity. Whether the poll indicated active support for maintaining soil productivity and a strong agriculture or a passive wish for the ultimate in quality of life remains to be seen. The necessity for a "nationally accepted, clearly defined set of objectives for soil conservation, especially for public programs in the field" is greater now than at any other time in recent history. Such objectives—or the lack of them—will profoundly affect soil productivity, the American public, and the world.

REFERENCES

Adams, W.E. *Journal of Water and Soil Conservation* 4(1949):130.

Adams, E.P.; Blake, G.R.; Martin, W.P.; and Boelter, D.H. *Transactions, 7th Congress of the Soil Science Institute*, Vol. 1, pp. 607-15. Madison, WI.: Soil Science Institute, 1961.

Allmaras, R.; Gupta, S.C.; Pikul, J.L., Jr.; and Johnson, C.E. *Journal of Soil and Water Conservation* 34(1979):85-90.

Barlow, T. In *Soil Conservation Policies; an Assessment*, pp. 128-132. Ankeny, IA.: Soil Conservation Society of America, 1979.

Barrows, H.L., and Kilmer, V.J. *Advances in Agronomy* 15(1963):303-16.

Batchelder, A.R., and Jones, J.N., Jr. *Agronomy Journal* 64(1972):648-82.

Berg, N.A. In *Soil Conservation Policies; an Assessment*, pp. 8-17. Ankeny, IA.: Soil Conservation Society of America, 1979.

Blake, G.R.; Nelson, W.W.; and Allmaras, R.R. *Proceedings, Soil Science Society of America* 40(1976):943-48.

Brady, N.C. *The Nature and Property of Soils*. New York, NY: Macmillan, 1974.

Bolton, E.F., and Aylesworth, J.W. *Canadian Journal of Soil Science* 39(1959):98-102.

Campbell, R.B.; Matheny, T.A.; Hunt, P.G.; and Gupta, S.C. *Journal of Soil and Water Conservation* 34(1979):83-5.

Council on Agricultural Science and Technology (CAST). Report No. 21. Ames, IA.: CAST, 1973.

Cory, D.C., and Timmons, J.F. *Journal of Soil and Water Conservation* 33(1978): 221-26.

Crosson, P.R. *Farmland, Food, and the Future*, pp. 99-111. Ankeny, IA.: Soil Conservation Society of America, 1979.

Doneen, L.D., and Henderson, D.W. *Agricultural Engineering* (1953):94-95;102.

Engelstad, D.P. and Schrader, W.D. *Proceedings, Soil Science Society of America* 25(1961 [a]).

Engelstad, D.P., and Schrader, W.D. *Proceedings, Soil Science Society of America* 25(1961 [b]):497-99.

Flocker, W.J.; Timm, H.; and Vomocil, J.A. *Agronomy Journal* 52(1960):345-48.

Frey, H.T. *Major Uses of Land in the U.S.: 1974*. U.S. Department of Agriculture, Economics, Statistics, and Cooperatives Service. Agricultural Economic Report No. 440. Washington, D.C.: USDA, 1979.

Gill, W.R. *Compaction of Agricultural Soils*, pp. 431-58. St. Joseph, MI.: American Society of Agricultural Engineers, 1971.

Greb, B.W. *Journal of Soil and Water Conservation* 34(1979):269-73.

Greb, B.W.; Smika, D.E.; and Welsh, J.R. *Journal of Soil and Water Conservation* 34(1979):264-68.

Gupta, S.C.; Onstad, C.A.; and Larson, W.E. *Journal of Soil and Water Conservation* 34(1979):77-79.

Hargett, N.L. and Berry, J.T. *1978 Fertilizer Summary*. Muscle Shoals, AL.: TVA, 1979.

Heady, O.E., and Timmons, J.F. *Journal of Soil and Water Conservation* 30(1975):15-22.
Held, R.B., and Clawson, M. *Soil Conservation in Perspective*. Baltimore, MD.: John Hopkins Press, 1965.
Hoeft, F.E. *Soil Erosion and Sedimentation*, pp. 23-30. 1977.
Hoeft, F.E. In *Soil Conservation Policies; and Assessment*, pp. 104-12. Ankeny, IA.: Soil Conservation Society of America, 1979.
Holt, R.F. *Journal of Soil and Water Conservation* 34(1979):96-98.
Koelsch, R.K.; Clark, S.J.; Johnson, W.H.; and Larson, G.H. In *Food, Fertilizer and Crop Residues*. Ann Arbor, MI.: Ann Arbor Scientific Publications, 1977.
Larson, W.E. *Journal of Water and Soil Conservation* 34(1979):74-6.
Langdale, G.W.; Box, J.E., Jr.; Leonard, R.A.; Barnett, A.P.; and Fleming, W.G. *Journal of Soil and Water Conservation* 34(1979):226-28.
Libby, L.W., and Birch, A. In *Soil Conservation Policies; an Assessment*, pp. 136-46. Ankeny, IA.: Soil Conservation Society of America, 1979.
Lindstrom, M.J.; Gupta, S.C.; Onstad, C.A.; Larson, W.E.; and Holt, R.F. *Journal of Soil and Water Conservation* 34(1979):80-2.
Lindstrom, M.J.; Skidmore, E.L.; Gupta, S.C.; and Onstad, C.A. *Journal of Environmental Quality* 8(1979):533-7.
Mahan, J.N., and Stroike, H.L. *The Fertilizer Supply*. U.S. Department of Agriculture, Agricultural Crop Stabilization Service. Washington, D.C.: USDA, 1978.
National Plant Food Institute. *Our Land and Its Care*. Washington, D.C.: NPFI, 1967.
New YorkTimes. 20 May 1979.
Olson, T.C. *Journal of Soil and Water Conservation* 32(1977):130-2.
Onstad, C.A., and Otterby, M.A. *Journal of Soil and Water Conservation* 34(1979): 94-6.
Otte, A.D., and LaConde, K.V. In *Food, Fertilizers and Agricultural Residues*, pp. 135-46. Ann Arbor, MI.: Ann Arbor Scientific Publications, 1977.
Phillips, R.E., and Kirkham, D. *Agronomy Journal* 54(1962):29-34.
Robertson, L.S., and Erickson, A.E. *Crops and Soils*, pp. 7-9. February, 1978.
Rosenberg, N.J. *Advances in Agronomy* 16(1964):181-96.
Sampson, R.N. *Journal of Soil and Water Conservation* 33(1978):206-08.
Skidmore, E.L.; Kumar, M.; and Larson, W.E. *Journal of Soil and Water Conservation* 34(1979):90-94.
Smith, F.W., and Cook, R.L. *Proceedings, Soil Science Society of America* 11(1946): 402-06.
Steffgen, F.W. *A New Look at Energy Sources*, pp. 23-36. Madison, WI.: American Society of Agronomy, 1974.
Timmons, J.F. In *Soil Conservation Policies; an Assessment*, pp. 53-74. Ankeny, IA.: Soil Conservation Society of America, 1979.
Thomas, B.J., and Cassel, D.K. *Journal of Soil and Water Conservation* 34(1979):20-24.
U.S. Department of Agriculture. "Potential Cropland Study." Washington, D.C.: USDA, 1975.
U.S. Department of Agriculture, Economics, Statistics, and Cooperatives Service. Statistical Bulletin No. 578. Washington, D.C.: USDA, 1977.
U.S. Department of Agriculture. *Agricultural Statistics, 1978*. Washington, D.C.: GPO, 1978.
U.S. Department of Agriculture, Economics, Statistics and Cooperatives Service. Statistical Bulletin No. 612. Washington, D.C.: USDA, 1978[a].
U.S. Department of Agriculture, Soil Conservation Service. *1977 National Resource Inventories*. Washington, D.C.: USDA, 1978[b].

U.S. Department of Agriculture. "Impact and Capability of Soil and Water Conservation Practices." Washington, D.C.: USDA, 1979.

U.S. Department of Agriculture. *Resources Conservation Act (RCA).* Washington, D.C.: 1979.

U.S. Department of Agriculture. Program Report and Environmental Impact Statement. Review Draft. Washington, D.C.: USDA, 1980.

Viets, F.J., Jr. *Fertilizer Use and Technology,* pp. 517-32. Madison, WI.: Soil Science Society of America, 1971.

Voorhees, W.B. *Journal of Soil and Water Conservation* 34(1979):184-87.

White-Stevens, R. In *Food, Fertilizers and Agricultural Residues,* pp. 5-26. Ann Arbor, MI.: Ann Arbor Scientific Publications, 1977.

Whitney, R.S.; Gardner, R.; and Robertson, D.W. *Agronomy Journal* 42(1950): 239-45.

Yeutter, C.K. *Journal of Soil and Water Conservation* 30(1975):12-4.

CHAPTER 3
Agricultural Research and the Future of American Agriculture

VERNON W. RUTTAN*

Prior to this century, world agricultural production was increased primarily by expanding the area cultivated, that is, by bringing new land into production. There were only a few exceptions to this generalization: in limited areas in East Asia, in the Middle East, and in Western Europe. By the end of this century, however, agricultural production will have to be increased primarily by higher yields, by increased output her hectare.

In most areas of the world, the transition from a resource-based to a science-based system of agriculture is occurring within a single century. In a few countries, including the United States, this transition began in the nineteenth century. In most of the presently developed countries, it did not begin until the first half of this century. Most of the countries of the developing world have been caught up in the transition only since mid-century.

In most developing countries, the institutional capacity to increase agricultural productivity at a rate consistent with current rates of growth in the demand for agricultural products has not yet been fully established. In developed countries, concern has shifted from the capacity to sustain growth in production to a concern with the design of policies and institutions to manage more effectively the use of agricultural technology.

In this paper, I first review the sources that have accounted for growth of agricultural production in the past. I then examine the more recent evidence on the contribution of research to growth in

*I am indebted to Sandra Batie, Robert Healy, Yao-chi Lu, and Luther Tweeten for comments on an earlier draft of this paper.

117

agricultural production. Finally, I present some of my own per-
spectives on issues related to the support for agricultural research
and the focus of agricultural research efforts over the next several
decades.

SOURCES OF GROWTH IN
AGRICULTURAL PRODUCTION*

During the remaining years of the twentieth century, it is imperative that both rich and poor countries design and implement more
effective policies to assure the growth of agricultural production.
A useful first step in thinking about this problem is to review the
approaches to agricultural development that have been employed
in the past and that will remain part of our intellectual framework
for thinking about agricultural development.

The literature on agricultural development can be characterized
as offering a half-dozen distinct explanations or "models" of agricultural development:

- The frontier model;
- The conservation model;
- The urban-industrial impact model;
- The diffusion model;
- The high-payoff input model;
- The induced innovation model.

These models should not be interpreted as sequential stages in
agricultural development. Rather they describe approaches that
have been and continue to be pursued, singly or in combination,
to increase agricultural production.

The Frontier Model

Throughout most of history, expansion of cultivated or grazed
areas has represented the dominant source of growth in agricultural production. The most dramatic example in Western history
was the opening up of the new continents—North and South
America and Australia—to European settlement during the eight-
eenth and nineteenth centuries. With the advent of cheap transportation during the latter half of the nineteenth century, the

---

*This section draws primarily on material originally presented in Hayami and
Ruttan (1971). It has been revised and edited in several more recent publications
(Binswanger and Ruttan, 1978; Yamada and Ruttan, 1980; Ruttan, Binswanger
and Hayami, 1980).

countries of the new continents became increasingly important sources of food and agricultural raw materials for the metropolitan countries of Western Europe.

In the United States, the potential for expansion of agricultural production by bringing new lands under cultivation was largely completed by the beginning of the twentieth century. The 1970s saw the "closing of the frontier" in most areas of Southeast Asia. In Latin America and Africa, the opening up of new lands awaits the development of technologies for controlling pests and diseases (such as the tsetse fly in Africa) or for releasing and maintaining the productivity of problem soils. By the end of this century, there will be few areas in the world where development along the lines of the frontier model will represent an efficient source of growth in agricultural production.

The Conservation Model

The conservation model of agricultural development evolved from the advances in crop and livestock husbandry associated with the English agricultural revolution and the notions of soil exhaustion suggested by the early German chemists and soil scientists. Until well into the twentieth century, the conservation model of agricultural development was the only approach to intensification of agricultural production that was available to most of the world's farmers.

The conservation model emphasizes the evolution of a sequence of increasingly complex land and labor-intensive cropping systems, the production and use of organic manures, and labor-intensive capital formation in the form of drainage, irrigation and other physical facilities to utilize land and water resources more effectively. The inputs used in the conservation system of farming—the plant nutrients, the animal power, land improvements, physical capital, and the agricultural labor force—are largely produced or supplied by the agricultural sector itself. Efforts to transplant the conservation model of agricultural development to the United States during the nineteenth century were frustrated largely by the high cost of labor and the low price of land. Initial success during the early decades of the twentieth century was reversed after 1940 by the sharp decline in the costs of energy used to produce machines, fuel, fertilizer, and pesticides.

The most serious effort to develop agriculture within the perspective of the conservation model in recent history was made by the People's Republic of China in the late 1950s and early 1960s. It became readily apparent, however, that the feasible growth rates, even under a rigorous recycling effort, were not compatible with modern growth rates in the demand for agricultural output, typically ranging from 3 to 5 percent in most less developed countries (LDCs). The conservation model remains an important source of productivity growth in most poor countries and an inspiration to agrarian fundamentalists and the organic farming movement in developed countries.

The Urban-Industrial Impact Model

In the conservation model, locational variations in agricultural development are related primarily to differences in environmental factors. This stands in sharp contrast to models that interpret geographic differences in the level and rate of economic development primarily in terms of the level and rate of urban-industrial development.

Initially, the urban-industrial impact model was formulated by von Thunen (in Germany) to explain geographic variations in the intensity of farming systems in the productivity of labor in an industrializing society. In the United States, the model was extended to explain the most effective performance of agriculture in regions characterized by rapid urban-industrial development, as opposed to regions where the urban economy had not made a transition to the industrial stage. In the 1950s, interest in the urban-industrial impact model reflected a concern with the failure of the agricultural resource development and price policies. These policies were adopted in the 1930s to remove the persistent regional disparities in agricultural productivity and in rural incomes in American agriculture.

The rationale for the urban-industrial impact model was to develop more effective input and output markets in areas of rapid urban-industrial development. Industrial development stimulated agricultural development by expanding the demand for farm products, by supplying the industrial inputs needed to improve agricultural productivity, and by drawing away surplus labor from agriculture. Empirical tests of the urban-industrial impact model have repeatedly confirmed that a strong nonfarm labor market is an essential prerequisite for growth of labor productivity in agriculture and improvement in the incomes of rural people.

The Diffusion Model

The diffusion of better husbandry practices was a major source of productivity growth even in pre-modern societies. The diffusion of crop and animals from the new world to the old—potatoes, maize, cassava, rubber—and from the old world to the new—sugar, wheat, and domestic livestock—was an important by-product of the voyages of discovery and trade from the fifteenth to the nineteenth centuries.

In the United States, the diffusion model has provided the major intellectual foundation of much of the research and extension effort in farm management and in rural sociology and economics since the emergence of these fields in the latter years of the nineteenth century. At that point, experiment station research was not yet capable of contributing significantly to agricultural productivity growth. Instead, emphasis was placed on transferring knowledge and technology from leading farmers to lagging farmers and from progressive areas to backward areas. In addition, the research of rural sociologists on the diffusion process contributed much to the effective diffusion of known technology.

The insights into the dynamics of the diffusion process, when coupled with the observation of wide agricultural productivity gaps among developed and less developed countries and a presumption of inefficient resource allocation among "irrational traditional-bound" peasants, produced an extension or a diffusion bias in the choice of agricultural development strategy in many less developed countries during the 1950s. During the 1960s, the limitations of the diffusion model as a foundation for the design of agricultural development policies became increasingly apparent as technical assistance and rural development programs, based explicitly or implicitly on the diffusion model, failed to generate either rapid modernization of traditional farms and communities or rapid growth in agricultural output.

The High-Payoff Input Model

The inadequacy of policies based on the conservation, urban-industrial impact, and diffusion models led, in the 1960s, to a new perspective—namely, that the key to transforming a traditional agricultural sector into a productive source of economic growth is investment designed to make modern high-payoff inputs available

to farmers in poor countries. With this model, peasants in traditional agricultural systems are viewed as rational, efficient resource allocators. This iconoclastic view was argued most vigorously by T.W. Schultz (1964). He insisted that peasants in
traditional societies remained poor because, in most poor countries, there were only limited technical and economic opportunities to which they could respond. The new, high-payoff inputs
were classified into three categories: (a) the capacity of public- and
private-sector research institutions to produce new technical
knowledge; (b) the capacity of the industrial sector to develop,
produce, and market new technical inputs; and (c) the capacity of
farmers to acquire new knowledge and use new inputs effectively.

The enthusiasm with which the high-payoff input model has
been accepted and translated into economic doctrine has been due,
in part, to the proliferation of studies reporting high rates of
return in the United States to public investment in agricultural
research and in the education of farm people. It also has been due
to the success of efforts to develop new high-productivity grain
varieties suitable for the tropics. New high-yielding wheat varieties were developed in Mexico, beginning in the 1950s, and new
high-yielding rice varieties were developed in the Philippines in
the 1960s. These varieties were highly responsive to industrial
inputs, such as fertilizer and other chemicals, and to more effective soil and water management. The high returns associated with
the adoption of the new varieties and the associated technical
inputs and management practices have led to rapid diffusion of
the new varieties among farmers in a number of countries in
Asia, Africa, and Latin America (Dalrymple, 1978).

An Induced Innovation Model

The high-payoff input model remains incomplete as a theory of
agricultural development. Typically, education and research are
public goods not traded through the marketplace. The mechanism
by which resources are allocated among education, research, and
other public- and private-sector economic activities was not fully
incorporated into the high-payoff model. The model does not
explain how economic conditions induce the development and
adoption of an efficient set of technologies for a particular society.
Nor does it attempt to specify the processes by which input and
product price relationships induce investment in research in a

direction consistent with a nation's particular resource endowments.

These limitations in the high-payoff input model led to efforts by Hayami and Ruttan (1971) to develop a model of agricultural development in which the appropriate path of technical change is determined by a nation's resource endowments. The induced innovation perspective was stimulated by historical evidence that agricultural technology is highly location-specific and that different countries had followed alternative paths of technical change in the process of agricultural development (Figure 1).

There is clear historical evidence that technology has been developed to facilitate the substitution of relatively abundant (hence cheap) factors for relatively scarce (hence expensive) factors of production. The constraints imposed on agricultural development by a relative scarcity of land have, in countries such as Japan and Taiwan, been offset by the development of high-yielding crop varieties designed to facilitate the substitution of fertilizer for land. The constraints imposed by a relative scarcity of labor, in countries such as the United States, Canada, and Australia, have been offset by technical advances leading to the substitution of animal and mechanical power for labor. In some cases, the new technologies—embodied in new crop varieties, new equipment, or

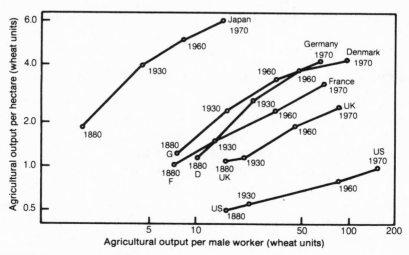

SOURCE: Binswanger and Ruttan, 1978

Figure 1. Historical Growth Paths of Agricultural Productivity in the United States, Japan, Germany, Denmark, France, and the United Kingdom, 1880-1970.

new production practices—may not always be substitutes for land
or labor by themselves. Rather, they may serve as catalysts to
facilitate the substitution of relatively abundant factors (such as
fertilizer or mineral fuels) for relatively scarce factors.

In agriculture, mechanical technology generally can be described as "labor-saving" while biological (or biological and chemical) technology is "land-saving." The primary effect of increased
mechanical technology is not to increase yields but to facilitate the
substitution of power and machinery for labor. Typically, this
results in a decline in labor per unit of land area. The substitution
of animal or mechanical power for human labor enables each
worker to extend efforts over a larger land area.

The primary effect of biological technology is to facilitate the
substitution of labor and/or industrial inputs for land. This may
occur through increased recycling of soil fertility by more labor-
intensive conservation systems; through the use of chemical fertilizers; and through husbandry practices, management systems,
and inputs (e.g., insecticides) that permit a more favorable production response to human effort.

Historically, there has been a close association between advances in output per unit of land area and advances in biological
technology, and between advances in output per worker and
advances in mechanical technology. These different productivity
growth patterns have given rise to the cross-sectional differences
in productivity and factor use illustrated in Figure 1.

Induced Technical Innovation in the United States and Japan

Induced technical change in agriculture can be seen more clearly
by drawing on the historical experience of the United States and
Japan. In the United States, it was primarily the progress of
mechanization, first using animals and later tractors for motive
power, which facilitated the expansion of agricultural production
and productivity by increasing the area operated per worker. In
Japan, it was primarily the progress of biological technology such
as higher yielding, more fertilizer-responsive crop varieties that
permitted rapid growth in agricultural output in spite of severe
constraints on the supply of land. These contrasting patterns of
productivity growth and factor use in United States and Japanese
agriculture can best be understood in terms of a process of
dynamic adjustment to changing relative resource endowments
and input prices.

In the United States, the long-term rise in wage rates relative to the prices of land and machinery encouraged the substitution of land and power for labor. This substitution generally involved the application of mechanical technology to agricultural production. The more intensive application of mechanical technology depended on the invention of technology that was more extensive in its use of equipment and land relative to labor. For example, the Hussey or McCormick reapers in use in the 1860s and 1870s required, over a harvest period of about two weeks, five workers and four horses to harvest 140 acres of wheat. When the binder was introduced, it was possible for a farmer to harvest the same acreage of wheat with two workers and four horses. The process illustrated by the substitution of the binder for the reaper has been continuous. As the limits to horse mechanization were reached in the early part of the twentieth century, the process was continued by the introduction of the tractor as the primary source of motive power. The process has been continued by the substitution of larger and higher-powered tractors and the development of self-propelled harvesting equipment.

In Japan, land was relatively scarce, and its price rose relative to wages. It was not, therefore, profitable to substitute power for labor. Instead, the new opportunities arising from the continuous decline in the price of fertilizer relative to the price of land were exploited through advances in biological technology. Crop variety improvement was directed, for example, toward the selection and breeding of more fertilizer-responsive varieties of rice. The enormous changes in fertilizer input per hectare that have occurred in Japan since 1880 reflect not only the effect of the response of farmers to lower fertilizer prices but also the development by the Japanese agricultural research system of "fertilizer-consuming" rice varieties to take advantage of the decline in the real price of fertilizer.

The effect of relative prices in the development and choice of technology is illustrated with remarkable clarity in the case of fertilizer in Figure 2, in which United States and Japanese data on the relationship between fertilizer input per hectare of arable land and the fertilizer-land price ratio are plotted for the period 1880-1960. In both 1880 and 1960, U.S. farmers were using less fertilizer than Japanese farmers. However, despite enormous differences in both physical and institutional resources, the relationship between these variables has been almost identical in the two countries. As the price of fertilizer declined relative to other fac-

tors, both Japanese and American scientists responded by investing crop varieties that were more responsive to fertilizer. However, American scientists always lagged by a few decades in the
process because the lower price of land relative to fertilizer
resulted in a lower priority being placed on yield-increasing
technology.

It is possible to illustrate the same process with cross-section
data in the case of mechanical technology. Variations in the level
of tractor horsepower per worker among countries largely reflect
the price of labor relative to the price of power. As wage rates
have risen in countries with small farms, such as Japan and Taiwan, it has been possible to adapt mechanical technology to the
size of the farm.

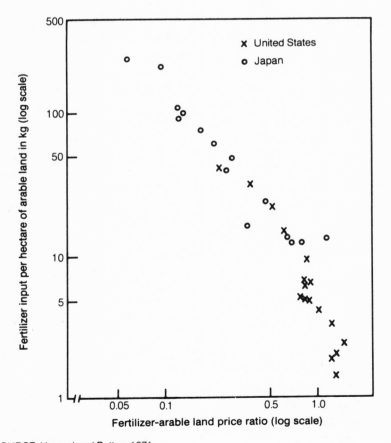

SOURCE: Hayami and Ruttan, 1971

**Figure 2. Relation between Fertilizer Input per Hectare of Arable
Land and the Fertilizer-Arable Land Price Ratio.**

The effect of a decline in the price of fertilizer relative to the price of land, or of the price of machinery and machinery services relative to the price of labor, has been to induce advances in biological and mechanical technology. The effect of the introduction of lower cost or more productive biological and mechanical technology has been to induce farmers to substitute fertilizer for land and mechanical power for labor. These reponses to differences in resource endowments among countries and to changes in resource endowments over time by agricultural research institutions, by the farm supply industries, and by the farmers have been remarkably similar in spite of differences in culture and tradition.

During the last two decades, as wage rates have risen rapidly in Japan and as land prices have risen in the United States, there has been a tendency for the pattern of technological change in the two countries to converge. During 1960s, fertilizer consumption per hectare rose more rapidly in the United States than in Japan, and tractor horsepower per worker rose more rapidly in Japan than in the United States. Both countries appear to be converging toward the European pattern of technical change in which increases in output per worker and increases in output per hectare occur at approximately equal rates.

There will be further changes in the future. During the 1970s, the price of energy rose. This has affected both the price of fuel and the price of fertilizer. It is unlikely that declining energy prices will, in the future, continue to determine the direction of scientific and technical effort in advancing either mechanical or biological technology.

THE CONTRIBUTION OF RESEARCH*

The beginning of successful modernization of agricultural production, as suggested in the previous section, is signaled by the emergence of sustained growth in productivity. During the initial stages of development, productivity growth usually is accounted for by improvement in a single partial productivity ratio such as output per unit of labor or output per unit of land. In the United States, and in other countries of recent settlement such as Canada, Australia, New Zealand, and Argentina, increases in labor productivity were the initial source of growth in total productivity. In countries that entered the development process with rela-

---

*This section draws very heavily on material presented in Evenson, Waggoner, and Ruttan (1979).

tively high labor-land ratios, such as Japan, Taiwan, Denmark, and Germany, increases in land productivity were initially the primary source of productivity growth.

As modernization has progressed, there has been a tendency for growth in total productivity—output per unit of total input— to be sustained by a more balanced combination of improvement in partial productivity ratios. Among the countries with the longest experience of agricultural growth, there tends to be a convergence in the patterns of productivity growth.

The changes in two partial productivity measures, land productivity and labor productivity, and in total productivity are illustrated for U.S. agriculture for the period 1950-1978 in Figure 3. During the 1950s and early 1960s, all three productivity measures grew rapidly. During the late 1960s, the rate of growth of land productivity and total productivity slowed down. During the 1970s, these two productivity indexes appear to have renewed their upward trend. Note also that the labor productivity index grew more rapidly than the total productivity index through the entire period. Part of the growth in labor productivity is due to higher capital investment per worker. The total productivity index grew at a slower rate because the services of the capital equipment, along with labor and other inputs, are included in the input index.

In Tables 1 and 2, changes in total productivity and in the two partial productivity growth rates are presented for the United States and Japan for the period since 1870. The tables illustrate the point made earlier that, prior to the mid-1950s, productivity growth in Japanese agriculture was dominated by growth in land productivity while prior to the 1940s, productivity growth in U.S. agriculture was dominated by growth in labor productivity.

The tables also show that both countries experienced periods of relatively slow productivity growth. During the first quarter of the twentieth century, the rate of growth in labor productivity declined in the United States. Total inputs grew more rapidly than output. Total productivity declined. Japan experienced its lowest rate of productivity growth during the period 1935-55.

Growth in total productivity has been influenced by a number of factors. Research leading to new knowledge and new technology has clearly been important. The education of farm people through formal schooling, organized extension activity, and agricultural publications has contributed to the rapid diffusion and efficient use of new technology. Transportation improvements

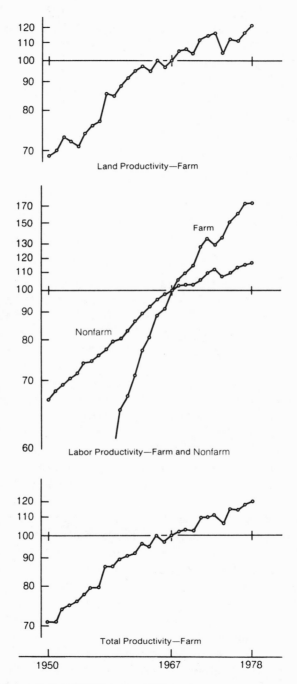

Land Productivity—Farm

Labor Productivity—Farm and Nonfarm

Total Productivity—Farm

SOURCE: Evenson, Ruttan, and Waggoner, 1979

Figure 3.   Productivity Measures (1967 = 100)

Table 1. Annual Average Rates of Change (Percent Per Year) in Total Outputs, Inputs, and Productivity in United States Agriculture, 1870-1979

Item	1870-1900	1900-1925	1925-1950	1950-1965	1965-1979
Farm Output	2.9	0.9	1.6	1.7	2.1
Total Inputs	1.9	1.1	0.2	-0.4	0.3
Total Productivity	1.0	-0.2	1.3	2.2	1.8
Labor Inputs	1.6	0.5	-1.7	-4.8	-3.8
Labor Productivity	1.3	0.4	3.3	6.6	6.0
Land Inputs	3.1	0.8	0.1	-0.9	0.9
Land Productivity	-0.2	0.0	1.4	2.6	1.2

SOURCE: Durost and Barton, 1960; Durost and Black, 1980

have reduced the costs of industrial inputs and the costs of marketing. Rural mail and telephone services have exerted a pervasive impact on productivity. The separate contributions of all of these factors have not yet been quantified. Considerable evidence has, however, been accumulated on the contribution of research and education.

The results of a large number of studies on the contribution of research to productivity growth have been assembled in Table 3. Almost all the studies indicate rates of return to investment in agricultural research well above the 10-15 percent rate (above inflation) usually considered adequate to attract investment. It is hard to imagine many investments in either private- or public-sector activity that would produce more favorable rates of return.

The contributions of research to increase agricultural productivity have been studied primarily by two methods: direct computation and regression analysis. The estimates listed under the "index number" heading in Table 3 were computed directly from the costs and benefits of research on, for example, hybrid corn. Benefits were estimated using accounting methods to measure the increase in production attributed to hybrid corn. The contribution of research was usually measured as the residual after all other factors that contributed to increased production were accounted for. The calculated returns represent the average rate of return per dollar invested over the period studied with the benefits of past research assumed to continue indefinitely. Benefits are defined as the benefits retained in the form of higher

Table 2. Average Annual Change in Total Outputs, Inputs, and Productivity in Japanese Agriculture, 1880-1975

Item	1880-1920	1920-1935	1935-1955	1955-1965	1965-1975
Farm Output	1.8	0.9	0.6	3.6	1.4
Total Inputs	0.5	0.5	1.2	0.7	—
Total Productivity	1.3	0.4	-0.6	2.9	—
Labor Inputs	-0.3	-0.2	0.6	-3.0	-3.6
Labor Productivity	2.1	1.1	0.0	6.6	5.0
Land Inputs	0.6	0.1	-0.1	0.1	-0.7
Land Productivity	1.2	0.8	0.7	3.5	2.1

SOURCE: Yamada, 1979

incomes to producers or passed on to consumers in the form of lower food prices.

The estimates listed under the "regression analysis" heading were computed by a different method, which permitted estimation of the incremental return from increased investment rather than the average return from all investment. This method can assign parts of the return to different sources, such as scientific research and extension advice. When regression methods are used, the significance of the estimated returns from research can be tested statistically. The dependent variable is the change in total productivity, and benefit is defined as the value of the change in productivity. The independent variables include research variables, which reflect the cost of research and the lag between investment and benefit. The objective of the regression procedure is to estimate that component of the change in productivity that can be attributed to research.

The effects of the timing and type of research have been analyzed in greater detail by Evenson (1978) for the United States.* These results, along with the regression equations used in the study, are presented in Table 4. The table shows that changes in the productivity of American agriculture from 1868 to 1971 were

---

*In the next several pages, I focus primarily on the results obtained by Evenson. A comparison with another important set of studies by researchers at Oklahoma State University and at the USDA is presented in Appendix A. The Oklahoma State-USDA studies on research productivity include Cline (1975); Lu, Cline and Quance (1979); and Knutson and Tweeten (1979). A similar model has been employed in a study by White, Havlicek, and Otto (1978).

Table 3: Summary Studies of Agricultural Research Productivity

Study	Country	Commodity	Time Period	Annual Internal Rate of Return (%)
Index number				
Griliches, 1958	USA	Hybrid corn	1940-55	35-40
Griliches, 1958	USA	Hybrid sorghum	1940-57	20
Peterson, 1967	USA	Poultry	1915-60	21-25
Evenson, 1969	South Africa	Sugarcane	1945-62	40
Ardito Barletta, 1970	Mexico	Wheat	1943-63	90
Ardito Barletta, 1970	Mexico	Maize	1943-63	35
Ayer, 1970	Brazil	Cotton	1924-67	77+
Schmitz & Seckler, 1970	USA	Tomato harvester	1958-69	
		with no compensation to displaced workers		37-46
		assuming compensation of displaced workers for 50% of earnings lost		16-28
Ayer & Schuh, 1972	Brazil	Cotton	1924-67	77-110
Hines, 1972	Peru	Maize	1954-67	35-40[a] 50-55[b]
Hayami & Akino, 1977	Japan	Rice	1915-50	25-27
Hayami & Akino, 1977	Japan	Rice	1930-61	73-75
Hertford, Ardila,	Colombia	Rice	1957-72	60-82
Rocha & Trujillo,	Colombia	Soybeans	1960-71	79-96
1977	Colombia	Wheat	1953-73	11-12
	Colombia	Cotton	1953-72	none
Pee, 1977	Malaysia	Rubber	1932-73	24
Peterson & Fitzharris,	USA	Aggregate	1937-42	50
1977			1947-52	51
			1957-62	49
			1957-72	34
Wennergren & Whitaker,	Bolivia	Sheep	1966-75	44.1
1977		Wheat	1966-75	-47.5
Pray, 1978	Punjab (British India)	Agricultural research and extension	1906-56	34-44
	Punjab (Pakistan)	Agricultural research and extension	1948-63	23-37
Scobie & Posada, 1978	Bolivia	Rice	1957-64	79-96
Regression analysis				
Tang, 1963	Japan	Aggregate	1880-1938	35
Griliches, 1964	USA	Aggregate	1949-59	35-40

Study	Country	Commodity	Time Period	Annual Internal Rate of Return (%)
Regression analysis (cont.)				
Latimer, 1964	USA	Aggregate	1949-59	not sig.
Peterson, 1967	USA	Poultry	1915-60	21
Evenson, 1968	USA	Aggregate	1949-59	47
Evenson, 1969	South Africa	Sugarcane	1945-58	40
Ardito Barletta, 1970	Mexico	Crops	1943-63	45-93
Duncan, 1972	Australia	Pasture improvement	1948-69	58-68
Evenson & Jha, 1973	India	Aggregate	1953-71	40
Cline, 1975 (revised by Knutson and Tweeten, 1979)	USA	Aggregate	1939-48	41-50[c]
		Research and extension	1949-58	39-47[c]
			1959-68	32-39[c]
			1969-72	28-35[c]
Bredahl & Peterson 1976	USA	Cash grains	1969	
		Poultry	1969	
		Dairy	1969	
		Livestock	1969	
Kahlon, Bal, Saxena & Jha, 1977	India	Aggregate	1960-61	63
Evenson & Flores, 1979	Asia (national)	Rice	1950-65	32-39
			1966-75	73-78
	Asia (international)	Rice	1966-75	74-102
Evenson, Flores, & Hayami, 1978	Tropics	Rice	1966-75	46-71
	Philippines	Rice	1966-75	75
Nagy & Furtan, 1978	Canada	Rapeseed	1960-75	95-110
Davis, 1979	USA	Aggregate	1949-59	66-100
			1964-74	37
Evenson, Ruttan, & Waggoner, 1979	USA	Aggregate	1868-1926	65
	USA	Technology-oriented	1927-50	95
	USA	Science-oriented	1927-50	110
		Science-oriented	1948-71	45
	USA-South	Technology-oriented	1948-71	130
	USA-North	Technology-oriented	1948-71	93
	USA-West	Technology-oriented	1948-71	95
	USA	Farm management research & agricultural extension	1948-71	110

Table 4. Estimated Impacts of Research and Extension Investments in U.S. Agriculture

Period and Subject	Annual Rate of Return (%)	Percent of Productivity Change Realized in the State Undertaking the Research
1868-1926		
All agricultural research	65	not estimated
1927-1950		
Agricultural research		
Technology-oriented	95	55
Science-oriented	110	33

SOURCE: Evenson, Ruttan, and Waggoner, 1979

The Regression equations, standard errors of parameters (in parentheses), coefficients of determination (adjusted for degree of freedom), and numbers of observations (N) are as follows:

1868-1926

(1) $P = 45.29 + .521 \; INV + .813 \; RES + 3.04 \; LANDQ$
$(.162) (.171) (23.38)$

$ R^2 = .634; \quad N = 40 \text{ years}$

1927-1950

(2) $LN(P) = 1.40 \; LN(INV) + .106 \; LN(TRES) + .0000053 \; LN(TRES)*(SRES)$
$(.24) (.106) (.0000033)$

$ R^2 = .503; \quad N = 24 \text{ years x 4 regions}$

1948-1971

(3) $LN(P) = .0331 \; LN(TRES\text{-}S) + .0119 \; LN(TRES\text{-}N) + .0187 \; LN(TRES\text{-}W)$
$(.0085) (.0085) (.0089)$

$ + .2061 \; LN(TREX)* \; SRES + .3540 \; LN(ED) - .0394 \; LN(EXT)$
$(.0710) (.0426) (.0097)$

$ - .0116 \; LN(EXT)*ED + .1821 \; LN(TRES)*EXT$
$(.0021) (.0230)$

$ R^2 = .569 \quad N = 23 \text{ years x 48 states}$

related to the research performed by the state agricultural experiment stations and the U.S. Department of Agriculture (USDA). The effects of agricultural extension and the schooling of farmers also were accounted for in the analysis.

Period and Subject	Annual Rate of Return (%)	Percent of Productivity Change Realized in the State Undertaking the Research
1948-1971		
Agricultural research		
Technology-oriented		
South	130	67
North	93	43
West	95	67
Science-oriented	45	32
Farm management and agricultural extension	110	100

(4) $LN(P) = .0299\ LN(TRES-S) + .0040\ LN(TRES-N) + .0113\ LN(TRES-W)$
    (.0090)   (.0090)   (.0090)

$+ .5639\ LN(TRES)*SRES + .5855\ LN(ED) - .02539\ LN(EXT)$
    (.0104)   (.0369)   (.0102)

$.0196\ LN(EXT)*ED + .1369\ LN(TRES)*EXT + .00148\ LN(TRES)* SUB$
    (.0021)   (.0044)   (.00017)

$R^2 = .595$ N = 23 years x 48 states

Each equation also included region and time period dummy variables. The 1948-71 equations also included a business cycle variable and a cross-sectional scaling variable.

Variables:

P	Total productivity index;
INV	Index of inventions;
RES	Stock of all agricultural research with time weights;
LAND	Land quality;
TRES	Stock of technology-oriented research with time and pervasiveness weights (S, W, N, for South, West, North)
SRES	Stock of science oriented research;
ED	Schooling of farm operators;
EXT	Extension and farm management research stocks: LN is natural logarithm;
	* indicates variables multiplied.

During the 1868-1926 period, an estimated 65 percent annual rate of return was realized on this research investment. Evenson divided the research that occurred from 1927 to 1950 into two types. The first he called "technology-oriented research (TRES),"

defined as research in which new technology was the primary
objective. This included plant breeding, agronomy, animal production, engineering, and farm management. The second type he
called "science-oriented research (SRES)." Its primary objective
was to answer scientific questions related to the production of
new technology. Science-oriented research included research in
phytopathology, soil science, botany, zoology, genetics, and plant
and animal physiology. The SRES analyzed here is limited to that
conducted in institutions such as the state experiment stations or
the USDA where it is closely associated with TRES. It is possible
that the results might not apply, or would apply with a longer
time lag, to SRES that is isolated by organizational or disciplinary
boundaries.

From 1927 to 1950, TRES yielded an annual rate of return of 95
percent. During the same 23 years, SRES yielded an even higher
return, 110 percent. The 1927-50 period was one of substantial
biological invention, exemplified by hybrid corn, improvements in
the nutrition of plants and animals, and advances in veterinary
medicine. It was also a period of rapid mechanization. It is important to notice in the equations in Table 4 that SRES does not have
a significant independent effect. The high payoff to SRES is
achieved only when it is directed toward increasing the productivity of TRES.

Research conducted in one state changes productivity in other
states. This is referred to as "spillover." For 1927 to 1950, it was
estimated that 55 percent of the change in productivity attributed
to TRES conducted within a typical state was realized within that
state. The remaining 45 percent was realized in other states with
similar soils and climate. The spillover from SRES was considerably greater. For 1948 to 1971, the observations for individual
states allowed still more detailed analysis. TRES continued to yield
returns of over 90 percent. The payoff to research was especially
high in the South, where research had lagged in earlier periods.

SRES from 1948 to 1971 remained profitable as it interacted
with TRES, but it was less profitable than during 1927 to 1950.
The decline in the rate of return to SRES, both absolutely and
relative to applied research, between 1927-50 and 1948-71 is difficult to interpret. One interpretation is that basic research has
been a less serious constraint on advances in applied research in
the more recent period than in the earlier period. A second interpretation is that there has been a lack of effective articulation
between basic and applied research—that either basic research has

not been adequately focused in areas that are relevant to applied research or that applied research has not drawn adequately on potentially useful basic research. The continued high rates of return to applied research would seem to support the first interpretation.

Evidence concerning the effects on productivity of schooling and extension advice also can be obtained from the equations used to estimate the results presented in Table 4. The schooling of farm operators had a strong positive effect. The effect of extension education and farm management advice is more complex. Its impact was strongest in those states with considerable technological research and on farmers with little schooling. The effect of these interactions, combined with the direct effects of extension, was positive.*

The effect on productivity of locating research at multiple substations within each state also was captured by the regression equations in Table 4. There has been considerable debate on how a shift in the distribution of scientists among the central state stations and substations would affect the productivity of technological research. In the regression equation, the fraction in the substations (SUB) is multiplied by TRES. The interaction was positive and significant, indicating that decentralization has beneficially affected the productivity of state research systems.

An important and somewhat unexpected inference from the several rates of return to agricultural research studies is that public-sector agricultural research has accounted for considerably less than half of the growth in agricultural productivity in recent decades. A 10 percent increase in public-sector expenditures for agricultural research appears to increase the agricultural productivity index by only about 0.3 to 0.6 percent. This is only about 1/4 of the productivity growth rate in recent years.

But if rates of return to research are as high as suggested in Table 3, why do ever larger increases in investment in agricultural research have so little leverage? The answer is found in a very substantial under-investment in agricultural research.** The total investment in agricultural research is so small relative to agricul-

---

*The contribution of extension to productivity growth has been analyzed in greater detail by Huffman (1978).

**For an examination of some of the factors which explain the continued under-investment in agricultural research in the United States, see Evenson, Waggoner and Ruttan (1979) and Ruttan (1980).

tural production that even investments with very high rates of
return (at present levels of investment) only modestly affect the
rate of growth of agricultural output and productivity.

Total public sector agricultural research expenditures are approximately $1.0 billion (Figure 4). Of this amount over 40 per-
cent is from state appropriations. Estimates of agriculturally
related research in the private-sector also falls in the $1.0 billion
range. However, about half of private-sector research is conducted
by or for the food industries and is not directed toward expanding agricultural production.

CAN PRODUCTIVITY GROWTH BE SUSTAINED?

Will investment in research be adequate to sustain output and
productivity growth in American agriculture in the future?
Before responding to this question, it will be useful to review the
record of output and productivity growth during the last several
decades (Table 1).

The rate of growth of agricultural output increased from an
annual rate of 1.7 percent in 1950-65 to an annual rate of 2.2
percent during 1965-79. The 1965-79 rate was the highest for any
sustained period since the turn of the century, and it was achieved
in spite of a decline in the rate of productivity growth. The annual
rate of total productivity growth declined from 2.2 percent in
1950-65 to 1.8 percent in 1965-79. In that same period, the rate of
increase in labor productivity declined from 6.6 percent to 6.0
percent and the rate of increase in land productivity declined from
2.6 percent to 1.2 percent. Rising real prices of agricultural commodities were able to draw additional resources into production,
thus permitting the rate of growth of output to rise in spite of a
decline in the rate of productivity growth.

The evidence on lagging productivity growth has focused considerable concern on whether the support for agricultural research
has been adequate to sustain future productivity growth. This
concern has been reinforced by limited growth of federal support
for agricultural research since the mid-1960s (Figure 5). Support
for agricultural research expanded rapidly between 1950 and
1965. Between 1965 and 1978, federal support for agricultural
research grew, in real terms, at 0.4 percent per year. However,
nonfederal support grew at an annual rate of 3.9 percent during
this latter period.

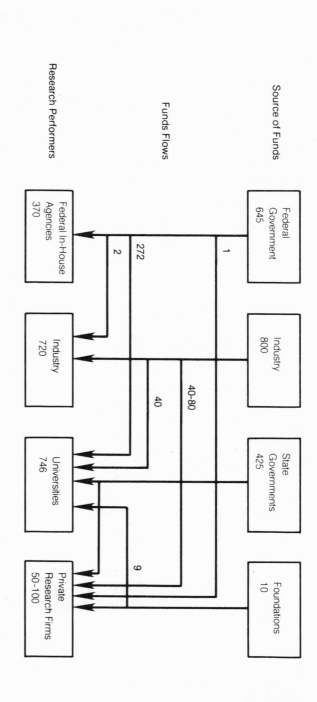

Source of Funds

Funds Flows

Research Performers

Figure 4. Research and Development Funds for the U.S. Food Research Systems, 1976 ($ million)

SOURCE: National Research Council, 1975b

This lag in the allocation of resources to research sharply contradicts the recommendations that emerged out of the intensive
joint USDA/State Experiment Station research planning effort in
1966. The projections presented in the National Program suggested the need for a 76 percent increase in scientific person-years
between 1965 and 1977. It also recommended a modest shift in
priorities from the commodity production, protection, and marketing categories toward the consumer protection and community development areas (Figure 6). During the projection period,
scientific person-years were reallocated among research program
areas roughly in line with the National Program recommenda-

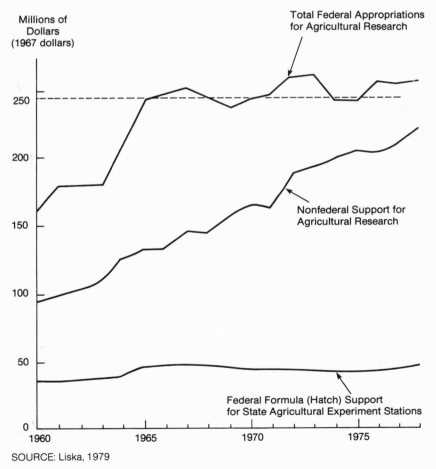

SOURCE: Liska, 1979

**Figure 5. Purchasing Power of Federal Appropriations and Non-
federal Support of Agricultural Research Programs in the U.S. for
1960-1978**

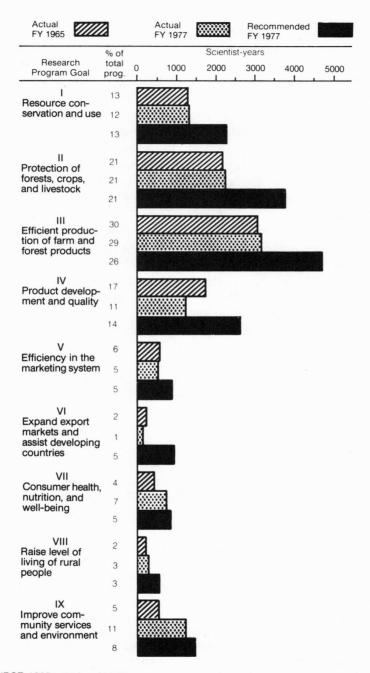

SOURCE: 1965 actual and 1977 recommended (USDA, 1966); 1977 actual (USDA, 1979)

Figure 6.   Scientist-years in SAES-USDA Program by Research Goal for FY 1965 and FY 1977 Compared to Recommended for 1977

tions. However, total scientific person-years devoted to agricultural research increased by less than 5 percent—from approximately 10,500 to just under 11,000. This overall increase conceals an actual decline in the number of USDA scientists which was slightly more than offset by an increase in the number of scientists at the state experiment stations.

The food crisis associated with the dramatic increase in grain imports by the USSR, the drought in the Sahel and in South Asia, and the sharp increase in petroleum prices in 1973-75 triggered a new set of evaluations of the adequacy of support for agricultural research (National Research Council, 1975a and b; Office of Technology Assessment, 1977). These studies had no more impact than the National Program in expanding research support, but they did result in a number of changes in the organization, administration, and funding of agricultural research at the federal level. One of these changes that has attracted considerable attention is the initiation of the USDA competitive grants program (Bredahl, Bryant and Ruttan 1980).

What can be concluded from the lag in research funding about the prospects of productivity growth in U.S. agriculture? Direct efforts to use historical research productivity estimates to project the effect of the future level of research support on productivity growth and on agricultural production capacity have been made by Lu, Cline, and Quance at the USDA (1979) (Table A1-2) and by Knutson and Tweeten (1979) at Oklahoma State University (Table A1-3). See appendix 1.

The USDA study projections, based on the historical model estimated by Cline for 1929-1972, were used to stimulate several scenarios for 1974-76 to 2000 and 2025. These results indicate the following.

> Under a low technology scenario in which nominal increases in public expenditures for agricultural R & E [research and education] are just offset by inflation, the annual growth rate in total productivity is 1 percent. Under a baseline scenario in which R & E grows 3 percent annually, the growth rate is 1.1 percent. The high technology scenario assumes that R & E grows 7 percent annually and that new and unprecedented agricultural technologies emerge as a consequence. The resulting growth rate is 1.3 percent. If the third scenario is projected to 2025 to allow more time for widespread adoption of new technologies, productivity can be expected to maintain the 1.5 percent historical growth rate of the past 50 years. (Lu, Cline, and Quance, 1979, p. 31)

The unprecedented new technologies that are built into the high technology scenario are photosynthesis enhancement, bioregulators, and induced twinning (in beef cattle). The effect of the technologies is to reduce the cost of achieving productivity growth. It is assumed that these new technologies will begin to have an impact on crop and animal production in the 1990s but that their major impact would be delayed until after 2000.

The projections developed by Knutson and Tweeten also are built on the model developed by Cline. Knutson and Tweeten developed three somewhat different scenarios. The *first* is a constant 3 percent per year real increase from 1976 to 2015; the *second*, a 10 percent annual increase from 1976-80 to compensate for the lag in research funding between 1966-76, followed by a 3 percent annual increase from 1981 to 2015; the *third* incorporates a 10 percent increase from 1976-80, followed by a 7 percent annual increase for 1981-2015 (Table A1-3). These projections suggest considerable difficulty in maintaining a productivity growth rate of 1 percent per year even after a significant "catch up" boost in research expenditures.

No attempt has been made to derive explicit projections based on Evenson's work. However, his results suggest somewhat greater leverage of research expenditures on productivity than is implied by the Oklahoma or USDA estimates.

In interpreting projections based on either the Oklahoma or USDA data or inferences about future productivity growth based on the Evenson analysis, several major qualifications should be observed. It was noted earlier that public-sector research accounted for only about 1/4 of productivity growth in the agricultural sector during 1950-1979. In both models, the increase in the educational level of farm people contributed even more significantly than research to productivity growth. My own guess is that improvements in the education of farm people will become a less important source of U.S. productivity growth in the future than in the past. A source of growth that is inadequately captured in both models has been the structural transformation of American agriculture—measured, but not fully captured, by the shift of labor out of agriculture and by the growth in farm size. My own guess is that structural change also will be a less powerful source of productivity growth in the future than in the past.

Another factor not adequately captured is the effects on productivity growth of private-sector research and development (R & D) and extension-type activities. Firm information on the expenditures and productivity for private-sector R & D is difficult to obtain. Estimates presented by the World Food and Nutrition Study (National Research Council, 1976b) and by the Agricultural Research Council (Williamson and Wilcke, 1977) suggest that expenditures on agricultural research and development in the private sector are at least as large as expenditures by the public sector (Figure 5). Private-sector agricultural research is much more heavily weighted toward the development end of the R & D spectrum; in some areas, such as research on pesticides and animal drugs, defensive research designed to secure or protect product registration has risen sharply during the last decade. In addition to the organized R & D efforts in the machinery, chemicals, and seed companies, the less formal developmental efforts by farmers, mechanics, and the smaller machinery firms that do not get reported as research and development expenditures continue to be an important source of advances in mechanical technology. My guess is that private sector R & D will become a larger source of productivity growth in agriculture during the next several decades. Much more careful analysis is needed of the organization and productivity of private sector R & D and of its articulation with public sector R & D. The Office of Technology Assessment (OTA) is currently reviewing the USDA proposals for reducing public-sector involvement in research and development on post-harvest technology and marketing. The USDA has taken the position that the reduced public-sector research and development in these areas will be compensated for by the private sector.

What implications can be drawn from the formal analysis and intuitive insights that are available to assess future rates of productivity growth? I find it hard to escape a conclusion that, unless the political and economic climate changes significantly, public-sector agricultural research expenditures in the United States in the immediate future will expand at considerably less than the annual real rate of 3 percent per year employed in the Oklahoma-USDA projections. Even when we attempt to account for the unaccounted, it is difficult to avoid concluding that the lag in research funding during 1965 to 80 will be followed by further declines in total productivity growth from 1980 through 2000.

I am skeptical, however, that we can expect to see a decline to 1 percent per year for a sustained period. At the same time, even a substantial effort to cash in on the higher rates of return available for agricultural research through rapid growth of research support will have great difficulty pushing the rate of productivity growth much above 1.5 percent per year. Even this would result in a continued decline from the 2.2 percent per year achieved during 1950-65 and the 1.8 percent achieved in 1965-79. This suggests a productivity growth rate more in line with 1925-50 than with 1950-80. It also suggests that prices of agricultural commodities will have to rise relative to the price of purchased inputs if output is to grow in the 1.5 to 1.6 percent annual range suggested in recent demand projections (White, Havlicek and Otto, 1978). Considerable caution is warranted, however, because of our limited capacity to project productivity growth rates over even the relatively short span of 20 years (see Appendix 2, which describes how far some past projections came from anticipating actual productivity change).

A PERSPECTIVE

I would now like to return to the implications of the induced innovation model outlined earlier in this paper. In retrospect, it appears that the major error in the resource and technology assessment studies of the early 1950s (Appendix 2) was a failure to understand the implications of declining real energy prices, particularly energy embodied in chemical inputs, as a focusing device for directing scientific and technical effort. As a result, the effects of the substantial interaction between advances in chemical and biological technology were underestimated.

There is now something approaching a consensus that the real price of energy embodied in agricultural inputs will rise in the future. Even those who resist this perspective do not expect real energy prices to decline over the next several decades. What will be the direction of technical effort induced by the changing input-input and input-product price relationships? My reading of the literature and sampling of scientific opinion suggest that we do not know. The closest analogy to the present situation in American agricultural history was the period between 1900 and 1925 (Table 1). With the closing of the frontier, productivity growth declined. The new sources of productivity growth, chemical and biological technology, did not begin to emerge for several decades. My guess is that it will be at least another decade before the direction of technical change induced by the rising real price of energy becomes clear.

The above perspective, if correct, has important implications for agricultural research management. Since we do not know where we are going, it is important that the exploration for new routes be kept as open as possible. Under these conditions, centralization of research management, particularly attempts to achieve a high degree of coordination among states and between the state and federal system, may come at a high price. This is a time to encourage parallel research and development efforts.* As the uncertainty increases, the value of redundancy rises. The historical evidence on research productivity suggests that a decentralized system more than compensates in productivity for the apparent losses due to redundancy. It is a time to avoid premature consensus on the opportunities that are ahead of us.

This places an extraordinary burden on research administrators in the states and in the USDA. They must go to the state legislature and the U.S. Congress with requests for expanded research resources. Yet there is no way that they can be confident of where the highest payoff will be to the research resources that become available. These judgments can only be made with any degree of authority by scientists who are on the leading edge of the individual disciplines and problem areas.

The evidence presented here also imposes a severe burden on the legislative bodies that provide the funding for public-sector agricultural research. The gains from agricultural research are realized with considerable time lag and over an extended period. This also means that the cost of current failure to fund agricultural research adequately, whether measured in terms of output and productivity growth, costs of production, food prices, or export earnings, will be felt only after considerable delay. Legislative bodies, like the rest of us, find it easier to deal with trade-offs between immediate short-run costs and benefits than between current costs and future benefits.

---

*For arguments which suggest the gains from parallel research and development efforts, see Nelson (1961) and Hirschman and Lindbloom (1962).

APPENDIX 1

The Oklahoma State-USDA Research Productivity Studies

The internal rates of return reported by Evenson are substantially higher than the rates reported in another series of important studies conducted at Oklahoma State University and at the USDA by Cline (1975), Lu, Cline, and Quance (1979) and by Knutson and Tweeten (1979). Some results from the Oklahoma State-USDA studies are summarized in Tables A1-1, A1-2, and A1-3.

At least part of the difference between the Evenson and the Oklahoma State-USDA estimates may be due to several of the differences in specification. The Evenson specification is more complete. In the Oklahoma-USDA study, a single rate of return is estimated for the combined effect of both research and extension. The Evenson results permit a separation of the effects of research and extension on productivity.

The Evenson results also employ a revision of the USDA productivity index, constructed by shifting factor weights annually (an approximation to the Divisa Index) rather than the periodic base period shift (the Laspeyres Index) employed by the USDA. As a result of this adjustment, the index constructed by Evenson rose more rapidly than the USDA index during the late 1960s. The effect of this revision is to increase the research coefficients in the Evenson estimates.

The coefficient for education (E) is higher in the Oklahoma-USDA specification than in the Evenson specification. This may be due, in part, to the inclusion of farm workers as well as farm operators in the Oklahoma-USDA specification. But it also seems likely that the Oklahoma-USDA estimates are picking up some of the effects of other omitted variables.

Table A1-1. Marginal International Rates of Return (%) to Production-oriented Research and Extension during Specified Time Periods

Period	13-Year-Lag	16-Year Lag
1939-48	40.9	49.7
1949-58	38.8	47.4
1959-68	31.6	39.4
1969-72	28.0	35.5

SOURCE: Knutson, 1979

The regression equation employed in estimating the above internal rates of return was:

$$1nP_t = \sum_{i=0}^{n} \beta_i 1nR_{t-1} + \beta_{n+1} \; 1nO_t + \beta_{n+2} \; 1nE_t + \beta_{n+3} \; 1nW_t + U_i$$

where

R_{t-1} = public sector production-oriented research and extension in preceding periods

O_t = public sector nonproduction-oriented research and extension in the present period

E_t = educational attainment farmers and farm laborers in current period

W_t = weather index for the current period

The results for the 13- and 16-year lag relatively were as follows:

	$1nR_{t-1} (\sum_{t-0}^{n} \beta_i)$	$1nE_t$	$1nW_t$	R_2	SEE[c]	DW[d]	ρ[e]
(1) 13 year:	.0369[a]	.7851 (3.0440)[b]	.0020 (4.7337)	.999	.02036	2.29	.839
(2) 16 year:	.0595[a]	.7299 (2.5554)	.0020 (4.3906)	.999	.02116	2.20	.819

a. A joint F test for each equation of the null hypothesis that all the regression coefficients for R's are equal to zero was rejected at the 1% level of significance in each case.

b. Numbers in parentheses are t-values.

c. Standard error of the estimate.

d. Durbin-Watson "d" statistic.

e. The estimated value of the first-order autoregression coefficient of the disturbances.

Table A1-2.   Annual Compound Rate of Growth in Agricultural
Productivity Under Alternative Rates of Investment in Research and
Extension and Growth of Education of Farmers and Farm Workers,
1980-2000.

	No Growth in R&E	Slow Growth in R&E	Rapid Growth in R&E	Rapid Growth in R&E plus Unprecedented New Technologies
Education	1.2	1.2	1.2	1.2
Research and Extension	0.0	3.0	7.0	7.0
Productivity	1.0	1.1	1.2	1.3

SOURCE: Lu, 1979

Table A1-3.   Annual Compound Rate of Growth (%) in U.S. Agricultural Productivity (Output per Unit of Conventional Inputs) under
Various Scenarious by Selected Time Periods, 1976-2015

Period	Scenario		
	T_3	$T_{10/3}$	$T_{10/7}$
1976-1985	1.036	1.102	1.115
1986-1995	.954	1.032	1.173
1996-2005	.866	.866	1.072
2006-2015	.801	.801	.986

SOURCE: Knutson, 1979

Note: Productivity growth was estimated with lag length of 16 years (see Table 6).

The increases in research expenditure projected in the three scenarios are as follows:

T_3	=	a constant 3% annual increase from 1976 to 2015
$T_{10/3}$	=	a 10% annual increase from 1976-80 followed by a 3% annual increase from 1981-2015.
$T_{10/7}$	=	a 10% annual increase from 1976-80 followed by a 7% annual increase from 1981-2015.

APPENDIX 2

A Retrospective View of Alternative Output, Input, and Productivity Projections, 1950-75

It may be useful to illustrate my caution about our capacity to project productivity growth rates by comparing a set of output/ input and productivity growth rate projections for 1960 and 1975 that I made in the mid-1950s, with changes that have occurred over the projection periods (Table A2-1). The projections were made to evaluate the implications of the projections of resource investment requirements being made by the USDA, the President's Water Policy Commission Report, and the President's Materials Policy Commission Report. These reports were concerned with the capacity of American agriculture to meet future food and fiber requirements and emphasized "the transitory nature of present food surpluses." Both reports projected substantial increases in land resource inputs to meet output requirements.

The approach employed in assessing these projections was to use an equation of the Cobb-Douglas type (linear in the logarithms), with a shift factor that captured the effect of productivity growth, to examine the consistency between the projected output requirements and alternative rates of growth of inputs and productivity. Four basic models with annual rates of productivity growth ranging from zero to 2.4 percent per year were calculated. The projections implied that continuation of even the relatively slow historical productivity growth rates could permit a relatively rapid growth in output with modest changes in land inputs (plus or minus 10 percent). The realized rate of productivity growth was, however, much higher than anticipated; it approached the most rapid rate projected. The other input projections were even less accurate. The decline in labor input was substantially underestimated in all models. And the current input levels that were actually realized were almost as high as those projected in the zero technical change model.

The projections in Table A2-1 were, of course, made at a time when the quantitative relationship between research investment and productivity growth was not as well understood as at present. Productivity accounting was a new craft. The Griliches study of the rate of return to investment in hybrid corn research (Table 3)

Table A2-1. Alternative Projections and Realized Farm Output and Factor Input Indexes for 1960 and 1975 (1950 = 100)

	Zero Technical Progress[a]		Slow Technical Progress[b]		Rapid Technical Progress[c]		Very Rapid Technical Progress[d]		Realized Technical Progress[f]
	Low land Inputs (I)	High land Inputs (II)	Low land Inputs (III)	High land Inputs (IV)	Low land Inputs (V)	High land Inputs (VI)	Low land Inputs (VII)	High land Inputs (VIII)	
1960 Projections Inputs:									
Labor	88	88	88	88	78	78	78	78	65
Land	96	104	96	104	96	104	96	104	90
Capital[e] (A)	178	172	140	136	149	143	124	121	114
(B)	183	177	145	140	153	147	127	124	
Current[e] (A)	214	207	169	163	178	172	148	145	170
(B)	204	198	161	155	171	164	141	138	
Contribution to output from:									
Inputs	122	122	112	112	110	110	100	100	96
Technological change	0	0	10	10	12	12	22	22	24
Total output	122	122	122	122	122	122	122	122	118
1975 Projections Inputs:									
Labor	81	81	81	81	67	67	67	67	30
Land	90	110	90	110	90	110	90	110	96
Capital[e] (A)	378	348	218	185	238	219	144	133	133
(B)	346	318	199	169	218	201	132	122	
Current[e] (A)	491	441	285	234	311	277	189	173	472
(B)	547	505	317	240	346	318	210	193	
Contribution to output from:									
Inputs	160	160	135	135	129	129	100	100	96
Technological change	0	0	25	25	31	31	60	60	54
Total output	160	160	160	160	160	160	160	160	150

SOURCE: Durost and Black, 1980.

a. Increased inputs are assumed to account for the entire increase in output.

b. Technological change is assumed to occur at a sufficiently rapid rate to permit an increase in output per unit of input of 1.0 percent per year between 1950 and 1975. This is the 1910-50 rate calculated on the basis of 1945-48 prices and techniques.

c. Technological change is assumed to occur at a sufficiently rapid rate to permit an increase in output per unit of input of 1.23 percent per year between 1950 and 1975. This is the 1910-14 rate calculated on the basis of 1910-14 prices and techniques.

d. Technological change is assumed to occur at a sufficiently rapid rate to account for the entire increase in output per unit of input. This requires an increase in output per unit of input of 2.2 percent per year between 1950 and 1960 and 2.4 percent per year between 1950 and 1975.

e. Estimate (A) for capital and current inputs is based on the assumption that the ratio of capital to current inputs (C_1/C_2) will continue to decline at the same percentage rate as during the period 1910-14 to 1945-48. Estimate (B) is based on the assumption that the 1925-27 to 1949-50 rate will continue. See text for further discussion of estimates A and B.

f. Calculated for 1948-53, 1958-63, and 1973-77. Capital indexes based on mechanical power and machinery; current inputs based on agricultural chemicals.

was several years in the future. Nevertheless, my cautious pessimism of the present may be only slightly more firmly grounded than my cautious optimism of the mid-1950s.

REFERENCES

Ayer, H.W. "The Costs, Returns and Effects of Agricultural Research in Sao Paulo, Brazil." Ph.D. dissertation, Purdue University, Layfayette, 1970.

Ayer, H.W., and Schuh, G.E. "Social Rates of Return and Other Aspects of Agricultural Research: The Case of Cotton Research in Sao Paulo, Brazil." *American Journal of Agricultural Economics* 54(1972):557-69.

Barletta, N.A. "Costs and Social Benefits of Agricultural Research in Mexico." Ph.D. dissertation, University of Chicago, 1970.

Binswanger, Hans P., and Ruttan, Vernon W. *Induced Innovation: Technology Institutions, and Development.* Baltimore: Johns Hopkins Press, 1978.

Bredahl, M., and Peterson, W. "The Productivity and Allocation of Research: U.S. Agricultural Experiment Stations." *American Journal of Agricultural Economics* 58(1976):684-692.

Bredahl, Maury E.; Bryant, Keith W.; and Ruttan, Vernon W. "Behavior and Productivity Implications of Institutional and Project Funding of Research." *American Journal of Agricultural Economics*, 1980.

Cline, Philip L. "Sources of Productivity Change in United States Agriculture." Ph.D. dissertation, Oklahoma State University, 1975.

Commission on International Relations, National Research Council. *World Food and Nutrition Study, Supporting Papers, Vol. V.* Agricultural Research Organization. Washington, D.C.: National Academy of Sciences, 1977.

Dalrymple, D.G. *Development and Spread of High-yielding Varieties of Wheat and Rice in Less Developed Nations.* Washington, D.C.: USDA, in cooperation with the U.S. Agency for International Development, 1978.

Davis, Jeff S. "Stability of the Research Production Coefficient for U.S. Agriculture." Unpublished Ph.D. thesis, University of Minnesota, 1979.

Duncan, R.C. "Evaluating Returns to Research in Pasture Improvement." *Australian Journal of Agricultural Economics* 16(1972):153-68.

Durost, D.D., and Barton, G.T. *Changing Sources of Farm Output.* USDA Production Research Report No. 36. Washington, D.C.: USDA, 1960.

Durost, D.D., and Black, T.E. *Changes in Farm Production and Efficiency.* USDA, Economics, Statistics, and Cooperatives Service. Statistical Bulletin No. 628. Washington, D.C.: USDA, 1980.

Evenson, Robert E. "The Contribution of Agricultural Research and Extension to Agricultural Production." Ph.D. dissertation, University of Chicago, 1968.

Evenson, Robert E. "International Transmission of Technology in Sugar Cane Production." New Haven, Conn., Yale University, 1969. Mimeographed.

Evenson, Robert E. "A Century of Agricultural Research, Invention, Extension and Productivity Change in U.S. Agriculture: A Historical Decomposition Analysis." Paper presented at Symposium on Agricultural Research and Extension Evaluation, 21-23 May 1978, Moscow, Idaho. Mimeographed.

Evenson, Robert E., and Flores, P. "Economic Consequences of New Rice Technology in Asia." International Rice Research Institute. Los Banos, Philippines, 1979.

Evenson, Robert E.; Flores, P.; and Hayami, Y. "Social Returns to Rice Research in the Philippines: Domestic Benefits and Foreign Spillover." *Economic Development and Cultural Change* 26(1978):591-607.

Evenson, Robert E., and Jha, D. "The Contribution of Agricultural Research Systems to Agricultural Production in India." *Indian Journal of Agricultural Economics* 28(1973):212-230.

Evenson, Robert E.; Ruttan, Vernon W.; and Waggoner, Paul E. "Economic Benefits from Research: An Example from Agriculture." *Science* 205 (1979): 1101-1107.

Griliches, Z. "Research Costs and Social Returns: Hybrid Corn and Related Innovations." *Journal of Political Economy* 66(1958):419-431.

Griliches, Z. "Research Expenditures, Education and the Aggregate Agricultural Production Function." *American Economic Review* 54(1964):961-74.

Hayami, Yujiro, and Akino, M. "Organization and Productivity of Agricultural Research Systems in Japan." In *Resource Allocation and Productivity*, edited by Thomas M. Arndt, Dana G. Dalrymple, and Vernon W. Ruttan, pp. 29-59. Minneapolis: University of Minnesota Press, 1977.

Hayami, Yujiro, and Ruttan, Vernon, W. *Agricultural Development: An International Perspective.* Baltimore: Johns Hopkins Press, 1971.

Hertford, R.; Ardila, J.; Rocha, A.; and Trujillo, G. "Productivity of Agricultural Research in Columbia." In *Resource Allocation and Productivity*, edited by Thomas M. Arndt, Dana G. Dalrymple, and Vernon W. Ruttan, pp. 86-123. Minneapolis: University of Minnesota Press, 1977.

Hines J. "The Utilization of Research for Development: Two Case Studies in Rural Modernization and Agriculture in Peru." Ph.D. dissertation, Princeton University, Princeton, N.J., 1972.

Hirschman, Albert O., and Lindbloom, Charles E. "Economic Development, Research and Development, Policy Making: Some Converging Views." *Behavorial Science* (1962):211-12.

Huffman, Wallace E. "Assessing Returns to Agricultural Extension." *American Journal of Agricultural Economics* 60(1978):969-75.

Kahlon, A.S.; Bal, H.K.; Saxena, P.N.; and Jha, D. "Returns to Investment in Research in India." In *Resource Allocation and Productivity*, edited by Thomas M. Arndt, Dana G. Dalrymple, and Vernon W. Ruttan, pp. 124-47. Minneapolis: University of Minnesota Press, 1977.

Knutson, Marlys, and Tweeten, Luther G. "Toward an Optimal Rate of Growth in Agricultural Production Research and Extension." *American Journal of Agricultural Economics* 61(1979):70-6.

Latimer, R. "Some Economic Aspects of Agricultural Research and Extension in the U.S." Ph.D. dissertation, Purdue University, Lafayette, 1964.

Liska, Bernard J., and Havlicek, Joseph, Jr. "Statement on Experiment Station Committee on Organization and Policy, Fiscal Year 1981 Budget Recommendations." Washington, D.C.: U.S. Department of Agriculture, Science and Education Administration, 1979.

Lu, Yao-chi; Cline, Philip; and Quance, Leroy. *Prospects for Productivity Growth in U.S. Agriculture.* USDA, Economics, Statistics, and Cooperatives Services. Agricultural Economic Report No. 435. Washington, D.C.: USDA, 1979.

Nagy, J.G., and Furton, W.H. "Economic Costs and Returns from Crop Development Research: The Case of Rapeseed Breeding in Canada." *Canadian Journal of Agricultural Economics* 26(1978):1-14.

National Research Council. *World Food and Nutrition Study: Interim Report.* Washington, D.C.: National Academy of Sciences, 1975 [a].

National Research Council. *World Food and Nutrition Study: Enhancement of Food Production for the U.S.* Washington, D.C.: National Academy of Sciences, 1975 [b].

Nelson, Richard R. "Uncertainty, Learning and the Economics of Parallel Research and Development Efforts." *The Review of Economics and Statistics* 43(1961):351-64.

Pee, T.Y. "Social Returns from Rubber Research on Peninsular Malaysia." Ph.D. dissertation, Michigan State University, 1977.

Peterson, W.L. "Returns to Poultry Research in the United States." *Journal of Farm Economics* 49(1967):656-69.

Peterson, W.L., and Fitzharris, J.C. "The Organization and Productivity of the Federal-State Research System in the United States." In *Resource Allocation and Productivity*, edited by Thomas M. Arndt, Dana G. Dalrymple, and Vernon W. Ruttan, pp. 60-85. Minneapolis: University of Minnesota Press, 1977.

Pray, C.E. "The Economics of Agricultural Research in British Punjab and Pakistani Punjab, 1905-1975." Ph.D. dissertation, Universtiy of Pennsylvania, 1978.

President's Material Policy Commission. *Resources for Freedom*. Washington, D.C.: USGPO, 1952.

President's Water Resources Policy Commission. *A Water Policy for the American People, Vol. I.*Washington, D.C.: USGPO, 1950.

Ruttan, Vernon W. "The Contribution of Technical Progress to Farm Output: 1950-75." *The Review of Economics and Statistics* 38(1956):61-4.

Ruttan, Vernon W.; Binswanger, Hans P.; and Hayami, Yujiro. "Induced Innovation in Agriculture." In *Natural Resources and Technological Progress*, edited by Mogans Boserup and Christopher Bliss, forthcoming, 1980.

Ruttan, Vernon, W. "Bureaucratic Productivity: The Case of Agricultural Research," *Public Choice*, 1980.

Schmitz, A., and Seckler, D. "Mechanized Agriculture and Social Welfare: The Case of the Tomato Harvester." *American Journal of Agricultural Economics* 52(1970):569-77.

Scobie, G.M., and Posada, T.R. "The Impact of Technical Change on Income Distribution: The Case of Rice in Columbia." *American Journal of Agricultural Economics* 60(1978):85-92.

Schultz, Theodore W. *Transforming Traditional Agriculture*. New Haven: Yale University Press, 1964.

Tang, A. "Research and Education in Japanese Agricultural Development." *Economic Studies Quarterly* 13(1963):27-41 and 91-9.

U.S. Congress Office of Technology Assessment. *Organizing and Financing Basic Research to Increase Food Production*. Washington, D.C.: USGPO, 1977.

U.S. Department of Agriculture and Association of State University and Land Grant Colleges. *A National Program of Research for Agriculture*. Washington, D.C.: USGPO, 1966.

U.S. Department of Agriculture. A National Program of Research for Agriculture. Washington, D.C.: USDA, 1966.

U.S. Department of Agriculture. Inventory of Agricultural Research, Fiscal Year 1977. Washington, D.C.: USDA Science and Education Administration, 1979.

Wennergren, E.B., and Whitaker, M.D. "Social Return to U.S. Technical Assistance in Bolivian Agriculture: The Case of Sheep and Wheat." *American Journal of Agricultural Economics* 59(1977):565-69.

White, Fred C.; Havlicek, Joseph, Jr.; and Otto, Daniel. "Agricultural Research and Extension Investment Needs and Growth in Agricultural Production." Blacksburg, Virginia: Virginia Polytechnic Institute and State University, 1978. Mimeographed.

Wilcke, H.L., and Williamson. *A Survey of U.S. Agricultural Research by Private Industry*. Washington, D.C.: Agricultural Research Institute, 1977.

Yamada, Saburo, and Hayami, Yujiro. "Agricultural Growth in Japan, 1880-1970." In *Agricultural Growth in Japan, Taiwan, Korea, and the Philippines*, edited by

Yujiro Hayami, Vernon W. Ruttan, and Herman Southworth. Honolulu: University Press of Hawaii, 1979.

Yamada, Saburo, and Ruttan, V.W. "International Comparisons of Productivity in Agriculture." In *New Developments in Productivity Measurement and Analysis.* Edited by John W. Kendrick and Beatrice N. Vaccara, pp. 509-594. Chicago: University of Chicago Press, 1980.

CHAPTER 4
Irrigation and the
Future of American Agriculture

KENNETH D. FREDERICK*

High-yield farming requires an adequate and timely supply of water. Over large areas of the western United States, rainfall is either insufficient or too unreliable to support a highly productive agriculture. In such areas, irrigation is essential for obtaining high crop yields. Even in some of the most humid areas, irrigation can help increase output by improving control over water, protecting against drought, facilitating double cropping, and encouraging the planting of higher-value crops.

Although irrigation's contribution to overall agricultural growth cannot be quantified with any precision, there are reasons to expect that the impact has been sizable in recent decades. Irrigated acreage has tripled since 1940 and doubled since 1950, periods of virtually no change in total cropland use. Furthermore, average yields and productivity have been greater on irrigated farms. In addition, within arid and semiarid areas, technological change has been higher on irrigated than on nonirrigated farms (U.S. Department of Agriculture [USDA], 1967, p. iv). All these factors have contributed to the increasing importance of irrigated agriculture, which currently accounts for more than 1/4 of the value of the nation's crops and nearly 1/7 of the nation's cropland.**

*Helpful comments on an earlier draft of the paper were received from Sandra Batie, Pierre Crosson, James Hanson, Robert Healy, and Leonard Shabman. Tables 4 and 5 were prepared by James Hanson. This paper draws on research partially financed by a grant from the Environmental Protection Agency's Environmental Research Laboratory in Athens, Georgia.

**According to estimates of the 1977 National Resource Inventory, cropland use included 55.8 million irrigated acres and 357.5 million dryland acres (USDA,

Irrigation can dramatically affect the productivity of arid and semiarid areas. For example, in southern California, seemingly commercially worthless desert lands have been turned into some of the the world's most productive agricultural areas through irrigation. In comparison, irrigation in the East has had less impact on yields in most years and is often used only with relatively high-value crops or to facilitate multiple cropping. In much of the East, the growth of irrigation has corresponded closely to changes in the demand for crops such as fruits and vegetables. Currently, irrigating rice and soybeans in the Mississippi Delta area and multiple cropping (such as corn or soybeans with a vegetable) in an area of south Georgia and northern Florida appear to be profitable at current price levels. Nevertheless, irrigation's impacts on the future of American agriculture will be determined largely by events in the West.

Western irrigation, which accounts for about 83 percent of the national irrigated acreage and produces more than 1/2 of the value of western crops, was spurred by the availability of low-cost water. Water initially was treated as a free good in the West. The original users not only were allowed to use the water without charge but were granted water rights for as long as the water was put to beneficial use. The only costs to the users were those associated with transporting the water from the source to the point of use. Moreover, federal irrigation projects provided large subsidies for the water development costs of about 1/5 of the West's irrigated acreage. Until recently, low energy prices also helped keep the costs of moving water low, a factor particularly important for the development of groundwater supplies.

The days of cheap water are ending in the West. In many areas, irrigators now depend on essentially nonrenewable water supplies. Current irrigation levels with average precipitation result in the mining of over 22 million acre-feet of water from western aquifers (U.S. Water Resources Council, 1978, Vol. 1, p. 18). The combination of declining groundwater tables and sharply rising energy prices is making groundwater pumping much more expensive. Surface water is not likely to pick up the slack, since the

---

1980). Data differentiating between the value of the production on irrigated and nonirrigated lands are not available for 1977. Lea, using a combination of data from the 1969 Census of Agriculture and 1971-73 estimates of the Statistical Reporting Service, estimated that 27 percent of the value of U.S. crop production was produced on irrigated lands (Lea, 1977, Table 2).

demands on the rivers and streams of the nation's principal irrigated areas already commonly exceed available supplies. Moreover, nonagricultural water demands are rising rapidly and undoubtedly will divert some water out of agriculture. Farmers throughout the West will find it increasingly difficult or expensive to maintain prior water withdrawal levels from many rivers, streams, and aquifers.

This paper will assess the impacts of water supplies on the future role of irrigation in American agriculture. The analysis focuses on the West, since this is where rising water costs and water shortages threaten not only expanded but also the current levels of irrigation. As background for assessing future changes in western irrigation, the paper examines (1) recent trends in irrigated agriculture, (2) the water supply situation, (3) likely changes in water costs and their impacts on the profitability of irrigation and the costs of alternative irrigation technologies, and (4) environmental factors affecting the long-term role of irrigation.

HISTORICAL TRENDS

Poor and contradictory data pose a major obstacle to determining both current and historical irrigation levels. Only two primary data sources provide estimates of irrigated acreage in every state: (1) the agricultural censuses taken about every five years (1974 is the most recent available), and (2) the 1967 and 1977 inventory data of the Soil Conservation Service (SCS). These SCS data are referred to as the 1967 Conservation Needs Inventory (CNI) and the 1977 National Resources Inventory (NRI). The 1974 Census of Agriculture shows only 41.2 million acres irrigated in the 50 states and 36.6 million in the 17 western states. In contrast, the 1977 NRI estimates irrigated acreage at 60.7 million in the 50 states and 50.2 million in the 17 western states. The 1977 national estimate is about 47 percent higher, and the western state estimate is 37 percent higher, than the 1974 census. The three years separating the census and the NRI estimates can explain only small portions of these differences. A detailed examination of the census, NRI, CNI, and other sources of irrigation data concludes: (1) the 1974 and 1969 agricultural censuses grossly understate the level of irrigation and (2) the statistical techniques underlying the 1977 NRI and 1967 CNI data combined with the reliance on trained personnel to gather and independently verify the data suggest that these estimates, especially on the national and regional levels, are the most reliable irrigation data available for all states.*

The agricultural censuses provide the longest series of data on irrigated acreage as well as the only data distinguishing between irrigated and dryland production and yields within all states. Consequently, some of the census data is used for this analysis. However, efforts are made to adjust for the deficiencies of the recent census data when these deficiencies appear significant.

The agricultural census data indicate a nearly continuous decline in both the annual rate of growth as well as the absolute growth in the number of acres irrigated in the West from 1947 to 1974 (see Table 1). For example, the 4.5 percent per annum growth of irrigated acreage and the average annual addition of nearly 1 million irrigated acres achieved from 1945 to 1950 were considerably higher than the growth achieved in any subsequent

Table 1. Growth of Irrigated Acreage in the West, 1945-1974

	Millions of Acres Irrigated	Change in Acres Irrigated[a]	Percent Change in Acres Irrigated	Average Annual Growth of Acres Irrigated (in percent)
1945	19.4			
		4.8	25	4.5
1950	24.3			
		2.7	11	2.7
1954	27.0			
		3.9	14	2.7
1959	30.7			
		2.5	8	1.6
1964	33.2			
		1.6	5	0.9
1969	34.8			
		1.9	5	1.1
1974	36.6			

SOURCE: U.S. Department of Commerce

a. Minor inconsistencies are due to rounding.

*Comparisons of the NRI and census data for several states for which independently gathered primary data are available support the view that the NRI data more accurately reflect irrigation levels. A detailed assessment of the irrigation data was undertaken as part of a large study of irrigation in the West which the author is undertaking with assistance from James Hanson. Examination of the data problems is contained in an appendix to the larger study, which is available from the author.

period. In contrast, irrigation grew only 1.1 percent per annum for an average of less than 400,000 acres per year from 1969 to 1974, according to census data.

While the rate of growth of irrigated acreage probably has declined from the rates achieved from 1947 to 1959, the changes implied by the 1967 CNI and the 1977 NRI data raise questions as to whether or not the absolute growth of irrigated acreage has actually declined. These sources show an 8.7 million acre increase or nearly a 2 percent annual growth rate from 1967 to 1977 in the 17 western states. This can be compared to the 3.4 million acre 10-year expansion implied by the 1964 and 1974 agricultural censuses. For reasons noted above, the higher increase is probably much closer to the actual growth.

These aggregate acreage figures mask significant regional changes. To understand the likely future of irrigation, it is desirable to focus on regions selected in part to reflect similar water supply and climatic conditions. Unfortunately, divisions along state boundaries are not adequate for this purpose. Table 2 uses an alternative breakdown of eight irrigation regions, with a further division of the High Plains region into three subregions. The regions are defined in the notes to the table.

Table 2 indicates the regional growth of irrigated acreage as well as the regional contribution to the overall expansion of western irrigation for the 1945 to 1974 period and three subperiods: 1945 to 1954, 1954 to 1964, and 1964 to 1974. A major consideration in making the regional divisions was to isolate the High Plains, an area covering parts of six states. There is no universally accepted definition of the High Plains. For this paper, the High Plains is defined as those parts of Colorado, Kansas, Nebraska, New Mexico, Oklahoma, and Texas that both overlie the Ogallala groundwater formation and fall south of the Platte River.* This definition, which excludes parts of Nebraska and the Dakotas commonly included in definitions of the High Plains, is adopted to focus on a region where: groundwater is virtually the sole water source; the aquifer is isolated from surrounding formations; and natural recharge is relatively insignificant. Thus, this definition

---

*Since the data are based on political and not hydrological boundaries, the area included does not conform exactly to our criteria. County-level data are required to estimate irrigation in these High Plains regions, and county estimates are available only in the agricultural censuses. Use of census data suggests the growth rates listed for the 1964 to 1974 decade may be low.

spotlights an important and extensive agricultural region where irrigation is resulting in large-scale mining of the water resources essential to continued irrigated farming. The severity of this regional problem is noted in the Second National Water Assessment, which cites groundwater overdrafts, estimated to be around 14 million acre-feet per year in the High Plains, as one of the nation's most critical water problems (U.S. Water Resources Council, 1978, Vol. 1, p. 78).

Additional reasons for according the High Plains special attention are suggested by the data in Table 2. The High Plains

Table 2. Growth of Irrigation by Irrigation Region

Irrigation Regions[a]	Average Annual Growth Rate of Irrigated Acres (percent/year)			
	1945-54	1954-64	1964-74	1945-74
(1) High Plains	20.0	5.2	2.9	8.7
(a) Southern	25.0	3.1	-0.3	8.2
(b) Central	28.8	12.8	5.6	14.5
(c) Northern	6.9	9.4	7.7	8.0
(2) Southern Plains	8.0	1.9	0	3.1
(3) Central Plains	7.4	6.5	6.0	6.6
(4) Dakotas	6.0	3.5	2.1	3.8
(5) Inter-Mountain	-0.1	1.4	-0.6	0.3
(6) Southwest	3.1	0	0.1	1.0
(7) California	4.0	0.8	0.2	1.6
(8) Northwest	3.6	2.0	0.4	1.9
Western States	3.7	2.1	1.0	2.2

SOURCE: U.S. Department of Commerce

a. Irrigation regions

(1) High Plains

    (a) Southern: Includes 33 Texas counties (Oldham, Potter, Carson, Gray, Wheeler, Deaf Smith, Randall, Armstrong, Donley, Parmer, Castro, Swisher, Briscoe, Bailey, Lamb, Hale, Floyd, Motely, Cochran, Hockly, Lubbock, Crosby, Yoakum, Terry, Lynn, Garza, Gaines, Dawson, Andrews, Martin, Howard, Midland, and Glasscock).

    (b) Central: Include 10 Texas counties (Dallam, Sherman, Hansford, Ochiltree, Lipscomb, Hartley, Moore, Hutchinson, Roberts, and Hemphill); 6 Oklahoma counties (Cimarron, Texas, Beaver, Harper, Ellis, and Woodward); 6 New Mexico counties (Lea, Roosevelt, Curry, Quay, Union, and Harding).

    (c) Northern: Includes 34 Kansas counties (Morton, Stevens, Seward, Meade, Clark, Stanton, Grant, Haskell, Gray, Ford, Hamilton, Kearny, Finney, Greeley, Wichita, Scott, Lane, Wal-

accounted for more than 4 of every 10 new acres irrigated in the West between 1945 and 1974; by 1974 this region had 22 percent of the total land irrigated in the West. Although the rate of growth as well as the annual change in the number of acres irrigated have declined sharply in the High Plains, this region's percentage contribution to the growth of western irrigation rose from about 38 percent prior to 1964 to 58 percent from 1964 to 1974.

Since there have been marked changes in the growth of irrigation within the High Plains, that area is further divided into three

Irrigation Regions[a]	Contribution to the Growth of Total Irrigated Acreage in the West (percent)			
	1945-54	1954-64	1964-74	1945-74
(1) High Plains	39	38	58	42
(a) Southern	33	16	-3	19
(b) Central	3	10	18	8
(c) Northern	3	13	42	14
(2) Southern Plains	12	6	0	8
(3) Central Plains	7	16	51	19
(4) Dakotas	1	1	1	1
(5) Inner-Mountain	-1	22	-17	4
(6) Southwest	6	0	0	2
(7) California	28	9	4	16
(8) Northwest	8	8	3	7
Western States	100	100	100	100

lace, Logan, Gove, Trego, Sherman, Thomas, Sheridan, Graham, Cheyenne, Rawlins, Decatur, Norton, Phillips, Kiowa, Hodgeman, Ness, and Rooks); 11 Colorado counties (Bace, Prowers, Kiowa, Cheyenne, Kit Carson, Yuma, Washington, Phillips, Logan, Sedgwick, and Lincoln); 9 Nebraska counties (Dundy, Chase, Perkins, Hayes, Hitchcock, Red Willow, Furnas, Frontier, and Harlan).

(2) Southern Plains: Texas and Oklahoma (excluding the High Plains).

(3) Central Plains: Kansas and Nebraska (excluding the High Plains).

(4) Dakotas: North and South Dakota.

(5) Inter-Mountain: Colorado (excluding the High Plains), Idaho, Montana, Nevada, Utah, and Wyoming.

(6) Southwest: Arizona and New Mexico (excluding the High Plains).

(7) California.

(8) Northwest: Oregon and Washington.

subregions. Early growth was heavily concentrated in the southern High Plains, i.e., Texas, south of the Canadian River. More recently, census data suggest irrigation in this region has declined slightly. While the average annual rate of growth in the central High Plains, i.e., west Texas north of the Canadian River, eastern New Mexico, and the Oklahoma Panhandle, has declined roughly by half every 10 years since 1945, the change in acres irrigated within this region as well as its percentage contribution to the total growth of irrigation in the West has increased. Since 1964, the most rapid growth of irrigation has been in the northern High Plains, i.e., western Kansas, eastern Colorado, and western Nebraska south of the Platte River, which accounted for 42 percent of the total growth of western irrigation from 1964 to 1974.

Overall, the growth centers of western irrigation have changed dramatically. From 1945 to 1954, 79 percent of the growth of irrigated acreage was in a southern belt that included the Southern Plains, the southern High Plains, the Southwest, and California. In subsequent decades, the contribution of this southern region to the overall growth of irrigation declined to 31 percent and then to 1 percent. In contrast, the Central Plains and northern High Plains combined contribution to the expansion of western irrigation rose from 10 percent over the 1945 to 1954 period to 29 percent over the 1954 to 1964 period and finally to 93 percent over the 1964 to 1974 period, according to census data.*

WATER AS A CONSTRAINT TO IRRIGATION

Regional changes in the expansion of irrigated lands can be explained in part by changes in the availability of water. As long as low-cost water was available, expansion was concentrated in (although certainly not limited to) the southern areas with the longer growing periods. Nebraska and Kansas became the centers of much of the new investment in irrigation only after the

---

*The contribution of the Central Plains and northern High Plains to overall growth in the 1964 to 1974 period probably is overstated by these numbers. According to census data, western irrigation outside the High and Central Plains declined between 1964 and 1974 due to a 600,000 acre decline in the Inter-Mountain region and only modest increases in other regions. Although the combination of the CNI and NRI data contradicts the conclusion that irrigation has declined outside the Central and High Plains, these data do confirm the conclusions drawn from the census data as to where most of the expansion has occurred.

demands on water started limiting expansion potential and raising water costs within the southern regions. Limited water supplies are likely to place increasingly tighter constraints on future irrigation. Consequently, a close look at water supplies and demands should help us understand both what has happened and what may happen with irrigated acreage.

The Bureau of Reclamation, recently renamed the Water and Power Resources Service, was established in 1902 to reclaim arid western land through irrigation. The Bureau helped ensure that the financial and engineering requirements of major water projects did not prevent the spread of irrigation. By 1975 Bureau projects were providing 26 million acre-feet of water annually for the irrigation of over 9 million acres. Over 4 million of these acres received all of their water from the Bureau; the rest of the acreage received supplementary water. In addition to providing financial and technical expertise and enabling farmers to benefit from the economies of scale associated with large-scale water projects, the Bureau further ensured the availability of inexpensive water for irrigation by providing direct subsidies. LeVeen (1978, pp. 2-3) estimates that at current collection rates and costs, the $3.62 billion the Bureau has spent for irrigation construction will be paid for from the following sources: approximately 56.7 percent from electricity sales, 40 percent from general tax revenues, and only 3.3 percent by the farmers.

Surface water withdrawals from western irrigation peaked around 88 million acre-feet in 1955. In subsequent years, irrigation withdrawals have fluctuated between 78 and 88 million acre-feet, reaching this higher level again only in 1978 (U.S. Department of Interior [USDI], 1951, 1957, 1961, 1968, 1972, 1977). This leveling off of streamflow diversions for irrigation reflects a number of factors, including the rising costs associated with developing new supplies, the competition from nonfarm users for underdeveloped supplies as well as water previously used for irrigation, and, as is demonstrated below, the relatively little surface water that remains to be developed within the areas with the most favorable growing conditions.*

---

*Changes in surface water withdrawals may not be directly related to changes in the acreage irrigated with surface water. For example, improvements in on-farm efficiency could increase the ratio of irrigated acreage to water withdrawn. However, since there has been little incentive for surface water users to improve irrigation efficiencies, it is not likely that there have

Groundwater started to become an important water source for irrigators in the 1930s. The combination of advances in engine and pumping technology and widespread drought encouraged the expansion of groundwater irrigation in the late 1930s. Additional technological advances as well as access to cheap energy for irrigation pumping further stimulated the growth of groundwater irrigation during the 1950s and 1960s. In particular, the availability of very low cost natural gas and development of the center pivot system stimulated the rapid growth of irrigation in the High Plains. The earliest available data on groundwater withdrawals for western irrigation indicate about 10.7 million acre-feet were withdrawn in 1945. Withdrawals rose more than five-fold to over 56 million acre-feet by 1975, with the greatest increases occurring in the Southern and Northern Plains (USDI, 1951, 1957, 1961, 1968, 1972, 1977 and USDI-Guyton).

The relative dependence on groundwater has increased markedly. From negligible levels in the early 1930s, groundwater rose to 21 percent by 1950 and 39 percent by 1975 of total withdrawals for irrigation. In 1975, the regional percentages were 78 percent in the Northern Plains and 80 percent in the Southern Plains, the two regions accounting for most of the net growth of irrigation over the past two decades.*

The expansion of groundwater use has led to mining of the aquifers underlying some important irrigation regions. In some cases, the depleted water tables already have forced a reduction in groundwater use. For example, in the Southern Plains, annual groundwater withdrawals for irrigation rose from less than 1 million acre-feet in 1945 to a peak of 13.3 million in 1965; declining groundwater tables were largely responsible for a subsequent decline in withdrawals to 9.6 million acre-feet in 1970 and 11.1 million in 1975.

Irrigation has placed enormous demands on western water; as of 1975, irrigators accounted for about 84 percent of the with-

---

been significant changes in this ratio. Furthermore, even if the ratio has increased, it does not necessarily follow that more land is irrigated with surface water. Improvements in on-farm efficiency reduce the runoff available for use downstream.

*The regions refer to the USDA farm production regions. The Northern Plains includes North and South Dakota, Nebraska, and Kansas; the Southern Plains includes Texas and Oklahoma. Water supply and use data cannot be broken down according to the irrigation regions defined in Table 2. Nevertheless, it is clear that the growth of groundwater use occurred principally in what Table 2 refers to as the High Plains and Central Plains.

drawals and 90 percent of the consumption. Although the West receives only about 27 percent of the annual rainfall of the lower 48 states, western fresh water consumption accounted for 84 percent of the national total and, on a per capita basis, was 14.7 times that in the East (USDI and USDA, 1977, pp. 30-31). As a result, western water supplies are seriously strained by current demands, and future demand will become increasingly competitive for limited supplies.

The normal measure of scarcity is the market price. But since water is seldom allocated within a competitive market, there are no simple measures of the seriousness of a region's water problems. Two indicators—(1) the relation between total water requirements and total streamflow, and (2) groundwater mining as a percentage of annual water consumption—are presented in Table 3 to indicate the pressures on water resources within water resource subregions.* Both indicators assume average year precipitation levels and 1975 consumption levels under average conditions.

Water requirements in Table 3 include 1975 offstream consumption requirements plus instream needs; instream needs are defined by whichever of the following water requirements is higher: navigational water requirements or the streamflow required to maintain habitat for aquatic and riparian plants and animals.** In an average year, water requirements exceed streamflows in 24 subregions, including about 66 percent of the West's irrigated land. In most other subregions, there is little slack between streamflows and requirements. In only 3 subregions with relatively poor agricultural potential are water requirements less than 75 percent of streamflows.

---

*The regions and subregions are defined according to drainage basins, and no combination of these conforms exactly to the boundaries of the 17 western states. Ten of these regions, which are divided into 53 subregions, include almost all of the area of the western states. But they also include 3 subregions that are either totally or largely within the eastern states and another 5 that include some parts of the East. The lands that are not common to both the 17 western states and water resource regions 9 to 18 are relatively insignificant in terms of irrigation, water use, and water supply problems. Consequently, there is little distortion introduced by using all 10 western water resource regions as a proxy for the 17 western states.

**The 1975 water requirements data of the Second National Water Assessment represent a consensus view of what requirements would have been in that year under average precipitation. For many subregions, the assumptions regarding the instream flows established for fish and wildlife are critical to the indicated degree of water scarcity.

Table 3. Indicators of the Demand Pressures on Western Water Resources in an Average Year

Region Number	Subregion Number	Name	Total Water Requirements as a % of Total Streamflow[a]	Groundwater Mining as a % of Annual Consumption
09		SOURIS-RED-RAINY	62	b
	01	Souris-Red-Rainy	62	b
10		MISSOURI	87	17
	01	Missouri-Milk-Saskatchewan	82	b
	02	Missouri-Marias	82	b
	03	Missouri-Musselshell	81	1
	04	Yellowstone	96	b
	05	Western Dakotas	84	2
	06	Eastern Dakotas	82	7
	07	North and South Platte	140	13
	08	Niobrara-Platte-Loup	103	13
	09	Middle Missouri	91	16
	10	Kansas	123	41
	11	Lower Missouri	87	5
11		ARKANSAS-WHITE-RED	83	68
	01	Upper White	84	2
	02	Upper Arkansas	134	3
	03	Arkansas-Cimarron	114	103
	04	Lower Arkansas	83	2
	05	Canadian	122	85
	06	Red-Washita	129	55
	07	Red-Sulphur	83	1
12		TEXAS-GULF	101	50
	01	Sabine-Neches	85	8
	02	Trinity-Galveston Bay	89	19
	03	Brazos	142	78
	04	Colorado (Texas)	119	38
	05	Nueces-Texas Coastal	96	26
13		RIO GRANDE	136	16
	01	Rio Grande Headwaters	110	b
	02	Middle Rio Grande	140	21
	03	Rio Grande-Pecos	148	46
	04	Upper Pecos	144	16
	05	Lower Rio Grande	136	1

The water supply situation obviously deteriorates when precipitation is less than average. Not only are supplies reduced but demand, especially irrigation demand, increases.* In a dry year, defined as a year that has less rainfall than at least 80 percent of

---

*Of all the demand uses considered, the Assessment estimates only irrigation and steam electric water uses according to the dryness of the year.

Region Number	Subregion Number	Name	Total Water Requirements as a % of Total Streamflow[a]	Groundwater Mining as a % of Annual Consumption
14		UPPER COLORADO	84	b
	01	Green-White-Yampa	87	b
	02	Colorado-Gunnison	80	b
	03	Colorado-San Juan	84	b
15		LOWER COLORADO	225	53
	01	Little Colorado	80	7
	02	Lower Colorado Main Stem	225	27
	03	Gila	304	61
16		GREAT BASIN	125	16
	01	Bear-Great Salt Lake	102	3
	02	Sevier Lake	186	40
	03	Humboldt-Tonopah Desert	177	27
	04	Central Lahontan	116	3
17		PACIFIC NORTHWEST	84	5
	01	Clark Fork-Kootenai	62	2
	02	Upper/Middle Columbia	79	8
	03	Upper/Central Snake	91	4
	04	Lower Snake	78	7
	05	Coast-Lower Columbia	85	2
	06	Puget Sound	81	b
	07	Oregon Closed Basin	101	2
18		CALIFORNIA	82	8
	01	Klamath-North Coastal	65	b
	02	Sacramento-Lahontan	76	4
	03	San Joaquin-Tulare	109	10
	04	San Francisco Bay	91	b
	05	Central California Coast	83	10
	06	Southern California	107	8
	07	Lahontan-South	243	43

SOURCE: U.S. Water Resources Council, 1978

a. Total water use includes instream use and consumption requirements. Consumption requirements are estimated as the total water that would have been consumed or lost assuming 1975
demand and precipitation levels that will be equaled or exceeded 50 percent of the time.
b. Indicates a percentage of less than 0.5.

all other years, water requirements exceed streamflows in 48 of
the 53 subregions. Four of the exceptions are in the well-watered
Pacific Northwest region; the other is the northern-most subregion in California. The magnitude of the dry-year supply short-
falls is particularly striking. In a dry year, in 26 of the subregions
requirements exceed streamflows by more than 50 percent, and
in 8 of the subregions requirements are more than twice dry-year

streamflows (U.S. Water Resources Council, 1978, Vol. 3, Appendix IV, Table IV-2).

Groundwater stocks permit a region's total water use to exceed its total streamflow in any given year. Indeed, there are major advantages in using groundwater to supplement surface water resources during unusually dry periods. Groundwater use can reduce the risk of substantial loss in a dry year and the need for costly surface storage, which also results in high evaporation losses. However, this is not the use to which western groundwater stocks generally have been put. Groundwater has become the principal source of supply in many areas, not simply a dry-year supplement. Consequently, enormous quantities of water are mined from western aquifers even in the wettest of years. In an average year, mining exceeds 20 billion gallons per day (U.S. Water Resources Council, 1978, Vol. 1, p. 18).

Groundwater mining is equivalent to 10 percent or more of total water consumption in 21 water resource subregions. Not surprisingly, there is a close parallel between these subregions and the 24 subregions where average-year water requirements exceed streamflow; 17 subregions belong to both groups, and these 17 account for about 86 percent of the total western groundwater depletions. Groundwater mining is 30 percent or more of consumption in 11 subregions and 50 percent or more in 5 subregions. The 5 subregions with the highest rates of groundwater mining include most of the High Plains of Texas, Oklahoma, New Mexico, Colorado, and Kansas, and Arizona's principal agricultural areas. These 5 subregions alone account for 20 percent of the West's irrigated lands. Despite the enormous quantities of water involved, groundwater mining does not threaten to exhaust the available water stored in any of the water resource regions or subregions during the foreseeable future. Groundwater stocks are large, and only in the California region does annual mining for an entire region approach 1 percent of available storage.* Of course, the percentages for individual subregions may exceed the regional levels. Nevertheless, mining is 1 percent or more of available storage in only 4 subregions—the North and South Platte River basin (subregion 1007) with 2.0 percent, the San Joaquin Valley (subregion 1802) with 1.8 percent, the central California

---

*Available storage is that portion of total groundwater storage that can be tapped with conventional wells, methods, and machinery.

coastal area (subregion 1805) with 1.0 percent, and the Arikaree-Republican River area of northern Kansas and southern Nebraska (subregion 1010) with 1.0 percent.

Aggregate data for regions as large and diverse as the water resource subregions are inadequate for analyzing the implications of groundwater mining for irrigated agriculture. The subregional data on groundwater mining as a percentage of available storage understate the problem facing many irrigators. Much of the water included as available storage underlies areas with low agricultural potential or is located at depths and in quantities that make it too costly for use in irrigation. Furthermore, seemingly modest reductions in the groundwater stocks of a large resource subregion obscure the damage experienced in a particular area when the reductions are localized.

Although irrigation demands continue to dominate water use, other water demands have grown faster in recent years and are expected to continue to grow faster. From 1960 to 1975, fresh water consumed by irrigators grew at only 1/2 the rate for all other users (USDI, 1960 and 1977). According to the average year projections of the Second National Water Assessment, western water consumption from 1975 to 2000 will increase only 6 percent for irrigation compared to 81 percent for all other users (U.S. Water Resources Council, 1978, Vol. 3, Appendix II, Table II-4). In the southern belt, where water is already very scarce,* irrigation water consumption is projected to decline by 1 percent from 1975 to 1985 and by an additional 3 percent from 1985 to 2000. In contrast, consumption for all other purposes is projected to rise by 54 percent by 2000. Since irrigation accounts for nearly 92 percent of 1975 water consumption within this water-scarce area, the large percentage increases in other uses and the modest declines in irrigation uses imply virtually no change in total water use within the 24 subregions where average-year water requirements exceed streamflow.

Development of the West's vast energy resources, especially coal and oil shale, is expected to place particularly heavy demands

---

*The water scarce area referred to includes the following 24 water resource subregions: 1007, 1008, 1010, 1102, 1103, 1105, 1106, 1203, 1204, 1205, 1301, 1302, 1303, 1304, 1305, 1502, 1503, 1601, 1602, 1603, 1604, 1803, 1806, and 1807. All these subregions have either high ratios of water requirements to streamflows or high ratios of groundwater mining to consumption; in most subregions, both conditions prevail. See Table 3.

on western water. The water consumption projections of the Second National Water Assessment include an allowance for steam electric production, petroleum refining, and fuels mining. Concern that the Assessment had not adequately accounted for all likely energy developments and associated water requirements led to a supplementary study by the Aerospace Corporation. The Aerospace Report (AR) accepts all the Assessment's water supply data and all the Assessment's demand projections except those relating to energy. From four energy development scenarios that are federally generated, the AR report determines the maximum feasible limits for energy development and the associated water requirements, assuming standard size plants and no special provisions to adopt water-conserving technologies. Consequently, the AR water-for-energy estimates are higher than any likely levels. The resulting nonirrigation water consumption levels in the AR estimates are 7 percent higher than the Assessment projections by 1985 and 39 percent higher by 2000.

In terms of total western water demand, the AR estimates represent a 1 percent increase by 1985 and 6 percent increase by 2000. Although these are only modest percentage changes in total western water demand, they will be localized and, within certain regions, may place major new demands on water supplies. Where demand already exceeds renewable supplies, any increased demand requires either reductions among other uses or additional groundwater mining. The actual impacts on irrigation will depend in large part on the institutions, including the legal system, affecting water use. These institutions vary widely among states.

While the struggle for western water is not new, it is likely to assume a new intensity in coming decades. The Second National Water Assessment concluded that almost every region west of the Mississippi River has insufficient water for agricultural production at present efficiency levels. Growth in water demand will aggravate the situation.

The earliest irrigators have a preferred position in this struggle for water since they commonly own the most senior water rights.* Potential new irrigators have no such legal advantage. In

---

*In several areas of the West, even the rights of some of the earliest irrigators are threatened by Indian water claims. Rulings of the nation's highest courts have consistently held that Indian lands have sufficient accompanying water rights to accomplish the purpose for which the lands were reserved and that these rights predate those of other users. Indian water rights have not been

fact, they face a major economic disadvantage in that irrigation generally is a relatively low-value user of water. For example, water costs of $100 per acre-foot or more are insignificant to the costs of mining or processing western coal, while water costs of half these levels are prohibitive for many irrigators, especially those growing relatively low-value crops. Thus, if market forces are allowed to operate, less water will be used for irrigation as demand starts to exceed an area's supply. Some farmers already have found it profitable to sell their water rights, and, unless they are legally constrained from doing so, many others are likely to follow suit.

Augmenting Water Supplies

Weather modification and interbasin transfers are two of the possible means of increasing water supplies in specific areas. Of the options, weather modification is more promising economically but less certain technically. Preliminary research suggests that winter cloud seeding under some conditions can augment water supplies. More specifically, studies suggest that a coordinated program to seed the clouds of six mountain ranges in the Upper Colorado River Basin can increase snowfall sufficiently to provide an additional 1.4 to 2.3 million acre-feet of water a year at a cost of about $5 to $10 per acre-foot. While this cost is attractive compared to most alternatives,* the technology is still unproved and likely to be limited in its transferability to other regions and seasons. Increasing rainfall through summer cloud seeding is a much tougher scientific problem than is winter cloud seeding. Although there has been some encouraging evidence from the seeding of summer clouds, the results have not been definitive, and it is difficult to know when a breakthrough might occur (Silverman, 1980).

Even if a weather modification technology is established and the economics are favorable, there will be major institutional obstacles to adopting a program. Paying for the program, compensating those that may be affected adversely by the program,

---

quantified, but they are potentially large and will necessitate reducing other uses. However, the impact on irrigation of satisfying Indian claims is not likely to be large, since irrigation is likely to be the primary use of Indian water.

*In California the dams and reservoirs under consideration will cost $200 to $300 per acre-foot of water added to the effective supply (Turner, 1980).

allocating the additional water, and eliminating the fears concerning adverse impacts on the weather downwind from the cloud seeding are difficult problems that likely will stall the adoption of weather modification long after scientists and economists are comfortable with the technology. The task of winter seeding is complicated since the losers (e.g., those with higher snow removal costs) are closer in both location and time to the seeding than are the beneficiaries (e.g., farmers perhaps hundreds of miles downstream who presumably receive more water many months later). Nevertheless, cloud seeding is used in a few western areas today, and its use is likely to spread as the value of water increases and the technology improves. Conservative estimates suggest that winter cloud seeding might add nearly 10 percent to the flow of the Upper Colorado River, which may be a reasonable target for the turn of the century. While this would represent a significiant increase for that area, it would be equivalent to only 0.3 percent of total western streamflows, and would have little impact on the overall course of western irrigation.

The scarcity of indigenous water resources in relation to demand has prompted a call for water imports from regions more amply endowed with water. Importing water has broad support within the lower Colorado River Basin and the High Plains areas where current activities depend on nonrenewable water sources. But if the net water supplies within the West's water-scarce southern belt are to be increased, the water will have to be transported over very long distances.* The Mississippi, Columbia, and Upper Missouri rivers are the closest from which large quantities might be obtained, although the availability of water for export would be contested vigorously in these areas. But even if there were plenty of water for export, the costs of transporting water to the High Plains or the Southwest would far exceed its value in agriculture. Although economic feasibility has not been a necessary requirement for western water projects, the economics of such a water transfer project appear poor even by the standards

---

*The Central Arizona Project, currently under construction, will make it possible to transport Colorado River water into the Phoenix-Tucson area. However, this project will not increase effective water supplies within the water-scarce areas of the West. The water is already being fully utilized; the project largely will divert water currently going to southern California to central Arizona.

of other water projects. For example, the value of the energy required to move water from the Mississippi to the Texas and New Mexico High Plains would be over $400 per acre-foot at 1980 electricity prices in the area.* This cost is well over 10 times the prices farmers say they can afford to pay for water and still irrigate profitably.

Future Water Use for Irrigation: Some Conclusions

The preceding examination of the use and supply of western water provides some insights for likely future changes in irrigation. Total surface water withdrawals for irrigation have fluctuated around a constant level since the mid-1950s. Two factors suggest this may continue for some time. Some farmers, particularly those within water-scarce areas, will have opportunities to sell water rights profitably. Since large percentage increases in other water uses can be accommodated with small changes in irrigation water use, the impacts on total irrigation will be gradual and not very large over the next several decades. Moreover, the transfers of surface water rights from irrigation to other uses are likely to be offset by modest increases in surface irrigation within the regions where water is still available for appropriation. The comparative stability of surface water use in the face of increasing water scarcity also reflects the insulation of most surface water costs from both market considerations and rising energy costs.

For several decades, the growth of western irrigation has been based on groundwater withdrawals, which rose threefold between 1950 and 1975. Future changes in irrigation are also likely to be determined largely by changes in groundwater irrigation, but the rate and even the direction of the net changes are uncertain. Groundwater uses have encountered sharply rising costs due to the combination of rising energy costs and declining groundwater tables. A look at the impact of these changes on water costs and the impact of higher water costs on irrigation techniques and the profitability of irrigating will provide a clearer view of future irrigation trends.

---

*A water import study by the Bureau of Reclamation in the early 1970s estimated that nearly 47 billion kilowatt hours (kwh) of power would be required to deliver 5,794,000 acre-feet of water to the High Plains annually. At 5 cents per kilowatt hour, approximately the average 1980 cost of electricity in Texas, the cost per acre-foot is over $400 (USDI, 1973).

CHANGING ECONOMICS OF
GROUNDWATER IRRIGATION

The overall impact of changing water supply conditions and
energy prices on western irrigation depends on the decisions of
thousands of farmers responding to their own distinctive investment opportunities. The relevant conditions, prices, and opportunities vary widely across and within regions. Farmers are
affected differently and can be expected to respond differently to
changing energy costs and groundwater conditions. Nevertheless,
the qualitative nature of the impacts are similar. Consequently,
insights into the likely direction and pace of future changes in
groundwater irrigation can be gained by illustrating how alternative energy prices and pumping depths affect water costs under
specific assumptions.

The impacts on water costs can be divided into two stages—the
costs of lifting water to the surface and the costs of distributing
the water to the plants. Table 4 indicates the cost of the energy
required to pump 1 acre-foot of water from various depths with
alternative fuels and fuel prices and a 60 percent pump efficiency.
The alternative energy prices represent actual levels in 1970 and
1980 and projections for 2000, all in 1977 constant dollars. The
range of alternative pumping depths—100, 200, and 300 feet—
includes the conditions facing many, but by no means all, farmers.
The table suggests that at least historically the type of fuel used
has been as important as the pumping depth and the price in
determining a farmer's energy costs.

Natural gas continues to be the cheapest means of pumping. At
1980 energy prices, these costs are only slightly more than 1/2 the
cost of the next cheapest alternative, electricity. Even when total
annual pumping costs are considered, natural gas is generally preferred, although the margin of its advantages over electricity is
narrowed considerably.* Except for farmers benefiting from very
inexpensive electricity, as in the Pacific Northwest, natural gas
would be the choice of virtually all groundwater users. However,

---

*Different fuels, of course, require different power units. Sheffield (Table 1)
compares the fixed costs of alternative power units to irrigate 130 acres with a
center pivot system. The annualized per acre costs are $6.10 for electric, $9.41
for natural gas, $9.39 for LPG, and $14.37 for diesel. The combined capital and
energy costs of a natural gas unit were less than those for electric. Total water
costs would include the cost of a pump and well, but these costs would be
comparable among the various fuels.

Table 4.   Energy Costs to Pump 1 Acre-Foot of Water from Different Depths with Alternative Fuels and Fuel Prices

(1977 current $)

	Pump Lift (ft.)	Energy Costs Under Alternative Fuel Prices		
		1970	1980	2000
Natural Gas	100	1.13	4.56	9.12
	200	2.30	9.29	18.58
	300	3.43	13.86	27.72
Electricity	100	7.52	8.88	17.76
	200	15.03	17.76	35.52
	300	22.55	26.64	53.28
Liquid Petroleum Gas (LPG)	100	7.32	12.60	25.20
	200	14.65	25.20	50.40
	300	21.98	37.80	75.59
Diesel	100	5.24	14.96	29.92
	200	10.59	30.00	60.00
	300	15.74	44.96	89.92

The technical assumptions such as the amount of fuel and the pressure required to lift an acre-foot of water are based on the study of Gordon Sloggett (U.S. Department of Agriculture, 1979). Other assumptions include a 60 percent pumping efficiency and the fuel costs in 1977 constant dollars. These costs are: natural gas ($/mcf) .39 in 1970, 1.58 in 1980, and 3.15 in 2000; electricity ($/kwh), .033 in 1970, .039 in 1980, and .078 in 2000; Liquid Petroleum Gas (LPG) ($/gal.) .25 in 1970, .43 in 1980, and .86 in 2000. The 1970 prices for LPG and diesel are national averages obtained from *Agricultural Prices, Annual Summary* (U.S. Department of Agriculture, June 1977). The 1970 price for electricity is from *Agricultural Prices* (U.S. Department of Agriculture, October 1977). There was a wide range in the natural gas prices paid by High Plains irrigators in 1970; the table assumes a price of 25c/mcf in 1970 dollars (Schwab). The 1980 prices reflect average prices paid by farmers in Nebraska, Kansas, Oklahoma, and Texas in January 1980. The prices were obtained through phone conversations with several officials in those states. Based on a conclusion of a recent major energy study (Landsberg, p. 71), it is assumed that real fuel prices will double by the year 2000.

natural gas has not been available to all farmers, and, as of 1977, only about 34 percent of the acreage irrigated with onfarm pumped water used natural gas.* This percentage is likely to decline, since farmers will find it increasingly difficult to arrange natural gas hookups for new irrigation. The resulting shift to more expensive fuels will tend to increase average water costs to new irrigators.

---

*About 49 percent of the acreage used electricity, 5 percent LPG, 10 percent diesel, and 1 percent gasoline (USDA, 1979, pp. 21-22).

Rising energy prices are changing the economics of irrigation, regardless of the fuel used. From 1970 to 1980, the real cost of pumping an acre-foot of water 200 feet rose an estimated $6.99 for natural gas, $2.73 for electricity, $10.55 for LPG, and $19.41 for diesel fuel. Cost increases are likely to be even higher over the next two decades. The assumptions of Table 4 suggest, for a 200-foot lift, additional increases exceeding $9 for natural gas, $17 for electricity, and $25 for both LPG and diesel. The greater the pumping lift, the higher the cost increases.

Once the water reaches the surface, additional energy is required to distribute the water to crops. The energy requirements vary widely, depending on the system used and the pressure at which it is operated. Table 5 illustrates the impacts of higher energy prices on the costs required to distribute 1 acre-foot of water with five different distribution systems. Surface irrigation requires very little energy for distribution, just enough to get the water to the highest point on the field so that gravity can take over. Center pivot and big gun systems, on the other hand, distribute the water under high pressure, which requires considerable energy. Of course, the operating costs of such systems are more susceptible than surface irrigation to changes in energy prices. For example, the difference between the electricity costs for distributing 1 acre-foot with a center pivot and a surface system rises from $11.28 at 1970 prices, to $13.33 at 1980 prices, and to $26.67 at prices projected for the year 2000.

Comparisons among irrigation systems should be based on the costs of delivering a given amount of water to the root zone of the plants. Surface irrigation tends to be considerably less efficient at this task than sprinkler and, especially, drip systems. To deliver 14 acre-inches of water to the root zone, a surface system with a 50 percent application efficiency requires pumping 28 acre-inches of water, while a center pivot system with an 80 percent efficiency requries pumping only 17.5 acre-inches of water. In this case, the difference between the total energy required to pump and distribute water with a center pivot versus a surface system depends on the pumping depth. Under current prices and the assumptions of Tables 4 and 5, the energy costs associated with a center pivot system drop below those of a surface system when the pumping lift rises to about 250 feet.

Farmers have a number of opportunities for offsetting some of the impacts of higher energy prices short of abandoning irrigation or switching to a completely different distribution system. A tail-water recovery system, which recycles the water that drains to the lower end of the field, can increase the application efficiency of surface distribution to levels comparable to those of sprinkler systems. This recovery system reduces the amount of water that must be pumped. Rising energy costs have made such recovery

Table 5.   Energy Costs to Distribute 1 Acre-Foot of Water per Acre with Different Distribution Systems and Alternative Fuels and Fuel Prices

(1977 current $)

Distribution System	Energy Costs under Alternative Fuel Prices		
	1970	1980	2000
Natural Gas			
Big Gun	4.36	17.60	35.20
Center Pivot	1.85	7.46	14.92
Other Sprinkler	1.06	4.26	8.52
Drip	.39	1.60	3.20
Surface	.13	.54	1.08
Electricity			
Big Gun	28.64	33.84	67.69
Center Pivot	12.15	14.36	28.72
Other Sprinkler	6.94	8.20	16.41
Drip	2.61	3.08	6.15
Surface	.87	1.03	2.05
LPG			
Big Gun	27.93	48.04	96.08
Center Pivot	11.85	20.38	40.76
Other Sprinkler	6.77	11.65	23.29
Drip	2.54	4.37	8.73
Surface	.84	1.46	2.91
Diesel			
Big Gun	20.01	57.16	114.31
Center Pivot	8.48	24.25	48.50
Other Sprinkler	4.85	13.86	27.71
Drip	1.82	5.20	10.39
Surface	.61	1.73	3.46

The energy cost estimates assume the following Pressure Per Square Inch (PSI) levels: 165 for big gun, 45 for other sprinkler, 5 for surface (U.S. Department of Agriculture, September 1979), 70 for center pivot (Sloggett), and 15 for trickle (Chen). Other assumptions are listed in the notes to Table 4.

systems economically attractive, and their use has spread rapidly over the last decade. The capital costs of the system can be recovered through energy savings in about 2.5 years with current electricity prices and a 200-foot pumping lift. Energy needs also are being reduced by lowering the pressure used with center pivot systems. Improving pump efficiency is still another means of reducing energy needs. Table 4 assumes a 60 percent pump efficiency, which is probably close to the average. Commonly, a $2,500 investment in new bowls and an engine tune-up are sufficient to improve pumping efficiency to about 75 percent. At current electricity prices and a 200-foot lift, the capital costs could be recovered from the energy savings in about 2 years. Irrigation scheduling to apply water only when needed by the crops may reduce the water applied without reducing yields. Services using computers to analyze weather, soil moisture, and plant water requirements can now be purchased in many areas for guidance in irrigation decisions.

To illustrate the impacts of rising energy costs and pumping lifts on profitability, farming in western Kansas, where irrigated corn and dryland wheat are the primary alternatives, may be considered. Based on 1977 crop budgets developed by the USDA, the returns to risk were $14.50 per acre for irrigated corn (assuming a yield of 104.2 bushels per acre, a price of $2.47 per bushel, pumping 2 acre-feet a height of 215 feet, a surface water distribution system, and natural gas for fuel costing $1.20/thousand cubic feet [mcf]) compared to $6.88 per acre for dryland wheat (assuming a yield of 25.3 bushels per acre and a price of $3.09 per 18 bushel).* If pumping depth increases by 2 feet per year, energy prices increase according to the assumptions of Table 4, and all other factors affecting profitability are unchanged, then the per acre returns to risk are reduced by 1980 to $1.35 for irrigated corn and $3.82 for dryland wheat and by 2000 to losses of $46.59 for irrigated corn and $3.32 for dryland wheat.

The above example overstates the likely impacts on irrigated profitability by assuming no adjustments to the rising water costs. Farmers already have made major adjustments to these and

---

*The returns to risk are net of the costs of management, general farm overhead, and rent. Higher energy prices also increase the cost of crop drying, which accounts for the decline in dryland profits. The assumptions are based on the budgets for the Kansas 100 region.

higher energy costs. Practices such as over-irrigating and lack of concern with the efficiency of either the pump or the distribution of the water, prevalent when water was cheap, are being phased out among many groundwater users. Also, cropping patterns have changed slowly, with the low-value crops that have high water requirements declining in importance. In the future, new technologies should provide farmers with a wider range of options for adjusting to high water costs. It has been only within the last decade that significant numbers of agricultural scientists have focused their efforts on developing more efficient irrigation systems, crop varieties with more efficient evapo-transpiration rates, and other technologies appropriate for a situation of high energy and water prices. These research efforts should provide irrigators with new techniques in the coming decades. Nevertheless, the available adjustments are not likely to alter the basic facts that rising water costs are reducing the profitability of irrigated farming relative to dryland farming and that the combination of higher energy costs and increasing water scarcity is limiting the opportunities for profitably expanding irrigated acreage in the West.

ENVIRONMENTAL CONCERNS

Failure to dispose adequately of the salts that inevitably emerge with irrigation reduces and, in extreme cases, may destroy the productivity of land. All water used for irrigation carries salts, and the salt content of the water rises as water evaporates, is consumed by plants, or passes over saline soils. If the salt concentrations become too high, they inhibit plant growth. Where drainage is adequate, salts can be flushed from the root zones if there is sufficient rainfall or if additional irrigation water is applied for this purpose. But these remedies seldom eliminate the salinity problem. When the salts are washed off the land with additional water applications, the water returning to a river or aquifer has even greater salt concentrations. This adversely affects downstream users and plants and wildlife dependent on the river. In areas lacking good drainage, repeated irrigations raise the level of the underground water table. If the water table reaches the root zone of the plants, capillary action carries water close to the soil surface, where it evaporates and leaves a salt residue. Over time, this process reduces and eventually destroys the productivity of the land.

Salinity is the most pervasive environmental problem stemming from irrigation in the United States. All western river basins
except the Columbia are confronted with high and generally rising salt levels. In much of the Colorado, parts of the Rio Grande,
and the western portion of the San Joaquin river basins, the salt
concentrations in either the water or the soils are approaching
levels that threaten the viability of irrigated agriculture. Groundwater salinity is not yet a widespread problem in the West. However, the problem is growing and already is serious in parts of
California, New Mexico, and Montana. Although groundwater
tables are less susceptible than surface waters to pollution, the
problems are much more difficult to correct underground (USDI,
1975, pp. 116-118).*

A rough but informed guess suggests that 25 to 35 percent of
the irrigated lands in the West have some type of salinity problem,
and the problems are getting worse (van Schilfgaarde, 1980). Two
large irrigated areas where salinity already is a serious problem
are the Lower Colorado River Basin and the west side of the San
Joaquin Valley in California. The underlying cause of the salinity
problem differs in each of these areas. In the Colorado, the salt
content of the river increases progressively downstream due to
the salt-concentrating effects of irrigation and the additional salts
picked up as the water passes other saline formations. About 2/3
of the salts delivered to the Colorado are from natural sources,
and these can only be eliminated through expensive investments
to divert the river and its tributaries around some of the areas
contributing the salts. Irrigation accounts for most of the remaining salts, which could be curtailed greatly by reducing irrigation
return flows. Ideally, water applications would be limited to the
quantities required by the plants and to what is needed to flush
the salts just beyond the root zone. But there is no incentive for
farmers to adopt such practices, since they derive no benefit from
using less water, and the resulting damages are passed downstream. Annual damages from salinity in the Colorado River have
been estimated between $75 and $104 million in 1980 and $122
and $165 million in 2000 if no control measures are taken. While
most of the damages would be incurred by municipal and indus-

---

*Groundwater salinity from irrigation may result from the infiltration of
irrigation waters or from the intrusion of sea water as fresh water is pumped
out of coastal aquifers.

trial users, farmers would face decreased crop yields, increased leaching requirements, and higher management costs. Four initial control projects costing a total of $125 million have been proposed (USDI, 1977).

The principal salinity problem in the San Joaquin Valley results from poor drainage, which prevents salt-laden waters from being carried away from the fields. High water tables already reduce the productivity of about 400,000 acres and threaten more than 1 million acres in the Valley. A $1.26 billion drainage system to carry irrigation runoff from the western side of the Valley to the Delta has been proposed as a solution to the problem. Since farmers would have to install their own underground drainage to get the waters to the central drain, the total costs of an effective system would be considerably higher than this figure (USDI, 1979).

Nonstructural measures might be undertaken to reduce salinity levels as well as the damages for any given level. Agricultural scientists have developed crops and irrigation techniques that enable farmers to irrigate successfully with high salt levels; more efficient onfarm water use can reduce the salt-loading impacts of irrigation. Van Schilfgaarde suggests that the San Joaquin Valley drainage problem could be reduced to about 5 percent of current levels and a long-term equilibrium reached through an integrated irrigation system whereby the best water is used first on salt-sensitive crops with the increasingly salt-laden runoff applied to increasingly salt-tolerant crops. The remaining high saline waters would be reduced to quantities that could be disposed of in evaporation ponds rather than requiring a costly transbasin drain as has been proposed.

As long as farmers pay only a small fraction of the social value of water, do not bear the costs of their own additions to salt loads, and expect the federal government to pay most of the costs of structural solutions, farmers responses to rising salinity levels will be largely to press for the government investments required to provide good quality water. When these efforts fail or when the government response lags, farmers will be forced to modify irrigation and cropping practices to cope with the salts. Undoubtedly, there will be some resulting decline in productivity and profitability. From a national perspective, basinwide irrigation management schemes probably offer the most efficient means of controlling salinity. However, the institutional obstacles to adopting such

schemes are enormous. Consequently, much greater emphasis is
likely to be placed on costly government projects than is required
or desirable.

Irrigation may have either positive or negative effects on erosion, a major environmental problem associated with agriculture.
On the positive side, irrigation helps provide a better vegetative
cover, which, in turn, reduces wind erosion and runoff. Also,
when land is leveled as part of an irrigation system, erosion is
reduced. From a national perspective, irrigation has expanded
greatly the land base suitable for high productivity agriculture,
and the high yields generally attained on irrigated lands enable
some highly erodable lands to remain under permanent vegetative cover. On the negative side, flood and furrow irrigation systems may cause serious erosion on lands that have not been
leveled adequately. Since about 70 percent of the West's irrigated
acreage uses flood or furrow distribution, the problem is at least
potentially widespread. Center pivot irrigation has led to cropping
hilly, sandy soils that are highly erodible, and the wheel tracks of
the large mobile sprinkler system have created even larger gullies.
Other than a few studies of individual irrigation projects, there
are no data to differentiate between the general erosion problem
and that resulting from irrigation.

Since much of the erosion as well as many other environmental
problems stemming from irrigation result from uncontrolled
water flows, there is considerable potential to control the problems through improved water management. Higher water costs
already have encouraged adoption of improved water management practices on many farms. The trend toward better water
management and the concomitant beneficial impacts on the
environment will continue in areas facing sharply rising water
costs.

THE FUTURE OF WESTERN IRRIGATION: SOME CONCLUSIONS

The overall growth of irrigation will depend largely on the real
price levels of agricultural products. Although some regions are
confronting physical constraints to expansion, there are additional
lands with access to water supplies, especially groundwater, that
can be irrigated. The important constraints for at least several
more decades will be economic, and these, of course, depend on
the relation between the prices of agricultural inputs and outputs.

Significant increases in product prices would erase the negative impacts of rising energy prices and pumping depths and would accelerate the development of new lands.*

Nevertheless, irrigation cannot be expected to continue to contribute to agricultural expansion the way it has in the past three to four decades. Surface water use for irrigation has been virtually level for several decades; future net changes will be modest and probably not very dependent on changes in price levels. No foreseeable crop price levels will make irrigation competitive with most municipal or industrial water uses. On the other hand, institutional factors favoring agricultural water uses, and the fact that small percentage changes in irrigation water use can accommodate large percentage increases in other uses, suggest that the negative impacts on surface irrigation from increasing water scarcity will be gradual. On balance, minor net declines in surface irrigation are likely as land and water move into other areas.

Groundwater irrigation will be more sensitive than surface irrigation to changes in agricultural crop price levels. In the absence of significant changes in crop prices, overall expansion may continue for another decade or so based largely on growth in the Nebraska sandhills. Irrigation is comparatively recent in this area, which has vast water resources, relatively low pumping depths, and low land prices. These features are attracting sizeable investments in center pivots. State irrigation experts estimate that Nebraska has a long-term sustainable physical limit of at least 15 million irrigated acres, an area twice that currently irrigated in Nebraska. They estimate that 10-12 million acres can be profitably irrigated at current prices (Supalla, 1980). Outside Nebraska, unless the impacts of rising pumping depths and energy prices are offset by higher crop prices, a net decline in groundwater irrigation is likely. On balance, expansion within Nebraska likely will provide for a slow but positive overall growth of western irrigation for another decade, even if energy price increases continue to outpace crop prices.

The long-term role of irrigation is threatened by resource and environmental factors. Perhaps 1/4 of those lands currently under irrigation are heavily dependent on nonrenewable water sup-

---

*For example, a 25 percent increase in the real price of corn would offset the production cost increases incurred between 1980 and 2000 on the hypothetical western Kansas farm considered earlier.

plies, and the productivity of several million additional acres is threatened by rising salt levels. The problems are spreading and increasing in intensity. Technology and the resulting changes in water use efficiency are important elements in adapting to these conditions. While new techniques for responding to higher water costs undoubtedly will be developed, much could be done now to improve water use efficiency. Improved management or new investments will provide opportunities for reducing water without reducing crop yields. Nevertheless, as water becomes more expensive, efforts to curtail its use may in some cases result in lower yields. Where water costs are high, farmers do adopt water conservation measures. But as long as legal and institutional factors insulate a large segment of irrigators from the changing resource conditions, inefficiencies will be common.

New water projects are not likely to contribute significantly to the future growth of western irrigation. Structural measures to develop new water sources, reduce salinity, or improve water distribution efficiency often are not cost effective when the water is to be used for agriculture. Most of the best water projects have been completed, and the changes in the criteria for evaluating new projects which took effect January 14, 1980, may make it more difficult to get new projects approved (U.S. Water Resources Council, 1979).

FURTHER THOUGHTS: THE NATIONAL IMPORTANCE OF IRRIGATION AND RESEARCH NEEDS

While there is widespread agreement that western irrigation is threatened by rising water costs and declining groundwater tables, there are widely divergent views on irrigation's importance to U.S. agriculture and the implications of any major reduction in irrigated acreage. Research done at Iowa State in the early 1970s downplays the importance of irrigation. For example, projections to 2000 suggest that if "allowed to be distributed in terms of competitive conditions and interregional comparative advantages (no supply control), the nation's domestic and export food demands could be met more economically by using less water in agriculture and by a wide redistribution of crop acreage" (Heady, 1976, p. 13). These projections are based on a national agricultural model that provides a least-cost solution for producing a given

level of agricultural output. Recent runs of an updated version of
the Iowa State model support the view that the nation's food and
fiber could be produced at lower cost with less irrigation. For
example, a July 21, 1979 model run shows only 33.4 million irrigated acres in the year 2000 with a moderate growth in demand
for agricultural products.*

Not surprisingly, this view is not shared by many whose livelihood depends on the output of irrigated lands. Great national and
global benefits in the form of abundant and cheap food and fiber
have been claimed to have resulted from the spread of irrigation
within the High Plains. And dire warnings have been raised about
the consequences of allowing that area to revert back to dryland
farming (Black, 1978, p. 147).

Self-interest undoubtedly underlies much of the concern for
the national implications of having irrigated acreage decline in an
area where groundwater stocks are being depleted. The absence
of sound analysis and reliable data undoubtedly have contributed
to and make it more difficult to respond effectively to the claims
about the benefits of irrigation. Indeed, less may be known about
the impact of irrigation on the overall performance of U.S. agriculture than is known about the impact of any of the other principal inputs. In addition, uncertainty as to the number of acres
irrigated contributes to our ignorance.

This paper suggests that irrigation currently is more extensive
than commonly believed and that, for several decades, irrigation
has played an important role in the growth of agricultural output
and productivity. But this paper also shows that the expansion
has been due in part to government subsidies for irrigation projects and has relied on the mining of groundwater. While these
factors have helped keep food prices down in times of scarcity,
agricultural surpluses and low farm prices have been more common problems during recent decades. During these periods, taxpayers have paid twice for the irrigation projects—first for the
direct subsidy and then for the crop price-support programs that
were required in part because of production of these subsidized
lands.

The costs to most groundwater users have been well below
social costs. In some cases farmers benefited from artifically low

---

*The July 21, 1979 model run was done for the USDA as part of that agency's
response to The Soil and Water Resources Conservation Act of 1977 (RCA).

energy prices. More generally, when groundwater tables are being depleted, a farmer's costs do not include the loss to either neighboring farmers or future users. Consequently, the groundwater is used at rates well in excess of socially efficient levels. While consumers may benefit in the short term from lower food prices, part of the costs are shifted to future generations. The inevitable adjustments to declining groundwater supplies will not be pleasant. Nevertheless, the socially most expensive response would be to provide subsidies either to enable farmers to pump to greater depths or to import water. An area where agriculture depends on declining groundwater supplies will inevitably become a higher cost producer. Spreading the cost increases among the general public clearly helps farmers in the affected region. But it also ensures higher overall production costs due to inefficient use of society's resources.

Western irrigation will not contribute to agricultural growth the way it did when water was abundant and cheap. However, if the transition from water abundance to scarcity allows for an efficient use of resources over time, irrigation will contribute to agricultural production and growth for many more decades. The serious problem will emerge if we attempt to keep water cheap when it is not. Such a policy will ensure its inefficient use and push the social costs of irrigated production to levels well above those of the dryland alternatives, thereby further limiting the long-term role of irrigation.

Water policy is dominated by entrenched beneficiaries of a system developed when water was plentiful in relation to demand. Although the underlying resource conditions as well as the national interests in the use of western water have changed dramatically within the last several decades, the laws and institutions that control and manage the resource have not been adjusted accordingly. The scientific community is proceeding with research that likely will provide farmers and others with more effective ways of responding to higher energy and water costs and the rising level of salinity. However, such developments will be adopted in a timely way only if the institutional factors provide the correct incentives. Herein lies the principal challenge to researchers and policymakers alike. There is a need to identify and develop strategies that will overcome the opposition to the reform of water development, allocation, and pricing policies and practices. In particular, the equity implications of alternative reforms and the possible measures for compensating those whose livelihood is affected adversely by such reforms need to be analyzed and taken into account in policy formulations if reform is to succeed.

REFERENCES

Aerospace Corporation. *Water Related Constraints on Energy Production.* Aerospace Report No. ATR-78 (9409)-1, June 1978.

Black, A.L. "High Plains Ogallala Study." In *The Multi-Faceted Water Crisis of West Texas,* edited by Frank L. Baird. Proceedings of a Symposium held 8-9 November 1978, Lubbock, Texas. Mimeographed.

Chen, Kuei-lin; Wensink, Robert; Wolfe, John. "A Model to Predict Total Energy Requirements and Economic Costs of Irrigation Systems." Presentation to the American Society of Agricultural Engineers, December 1976, Chicago, Illinois. Mimeographed.

Heady, Earl O. "U.S. Supply Situation for Food and Fiber and the Role of Irrigated Agriculture." Paper prepared for Texas A&M Conference, 25-26 March 1976, College Station, Texas. Mimeographed.

Landsberg, Hans H., *et al.* Energy: The Next Twenty Years. Cambridge, Massachusetts.: Ballinger, 1979.

Lea, Dallas. "Irrigated Agriculture: Past Trends, Present State, and Problems of Future Expansion." Draft, 1977. Mimeographed.

LeVeen, E. Phillip. "Reclamation Policy at a Crossroads." *Public Affairs Report,* vol. 19, no. 5. Bulletin of the Institute of Governmental Studies, University of California, Berkeley. 1978.

Supalla, Raymond; Johnson, Bruce; Axthelm, Leon; *et al.* University of Nebraska. Personal communication regarding irrigation issues in Nebraska. 1980.

Schwab, Delbert. Oklahoma State University, Stillwater, Oklahoma. Personal communication with James Pagana regarding natural gas prices.

Sheffield, Leslie F. "Energy—Is it the Achilles Heel for Irrigated Agriculture?" *Irrigation Age* (September, 1979).

Silverman, Bernie. Chief Atmospheric Water Resources Management, Water and Power Resources Service, Denver, Colorado. Personal communication, February 1980.

Sloggett, Gordon. Oklahoma State University, Stillwater, Oklahoma. Personal communication with James Hanson regarding irrigation distribution systems, 1980.

Turner, Kenneth M. Irrigation Specialist, Department of Water Resources, The Resources Agency, State of California. Personal communication, February 1980.

U.S. Department of Agriculture, Economic Research Service. *Measuring the Effect of Irrigation on the Rate of Technological Change.* Agriculture Economic Report No. 125. Washington, D.C.: USDA, 1967.

U.S. Department of Agriculture. *Basic Statistics—National Inventory of Soil and Water Conservation Needs, 1967.* Statistical Bulletin No. 461. Washington, D.C.: USDA, 1971.

U.S. Department of Agriculture, Economics, Statistics, and Cooperatives Service. *Agricultural Prices: Annual Summary, June 1977.* Washington, D.C.: USDA, 1977.

U.S. Department of Agriculture, Economics, Statistics, and Cooperatives Service. *Agricultural Prices, October 1977.* Washington, D.C.: USDA, 1977.

U.S. Department of Agriculture, Economics, Statistics and Cooperatives Service. Energy and U.S. Agriculture: Irrigation Pumping, 1974-77. Written by

Gordon Sloggett. Agricultural Economic Report No. 436. Washington, D.C.: USDA, 1979.

U.S. Department of Agriculture, Economics, Statistics and Cooperatives Service, in cooperation with Oklahoma State University. "Firm Enterprise Data Systems, 1974-1977." Mimeographed, 1976, 1977, 1978.

U.S. Department of Agriculture, Soil Conservation Service. "Basic Statistics— 1977 Natural Resources Inventory (NRI)." Revised February 1980. Mimeographed.

U.S. Department of Agriculture and U.S. Department of Interior. *Final Environmental Statement: Colorado River Water Quality Improvement Program, Vol. I, May 1977.* Washington, D.C.: USDA & USDI. 1977.

U.S. Department of Commerce. Bureau of the Census. *Census of Agriculture, 1945, 1950, 1954, 1959, 1964, 1969, 1974.* Washington, D.C.: U.S. Department of Commerce, 1947, 1952, 1956, 1961, 1967, 1973, 1977.

U.S. Department of Interior, U.S. Geological Survey. "Estimated Use of Ground Water in the United States—1945." Written by W.E. Guyton. Mimeographed.

U.S. Department of Interior, U.S. Geological Survey. *Estimated Use of Water in the United States in 1950.* Written by K.A. Mackchan.. U.S. Geological Survey Circular 115. Washington, D.C.: USDI, 1951.

U.S. Department of Interior, U.S. Geological Survey. *Estimated Use of Water in the United States in 1955.* Written by K.A. Mackchan. U.S. Geological Survey Circular 398. Washington, D.C.: USDI, 1957.

U.S. Department of Interior, U.S. Geological Survey. *Estimated Use of Water in the United States in 1960.* Written by J.C. Kammerer and K.A. Mackchan. U.S. Geological Survey Circular 456. Washington, D.C.: USDI, 1961.

U.S. Department of Interior, U.S. Geological Survey. *Estimated Use of Water in the United States in 1965.* Written by Richard C. Murray. U.S. Geological Survey Circular 556. Washington, D.C.: USDI, 1968.

U.S. Department of Interior, U.S. Geological Survey. *Estimated Use of Water in the United States in 1970.* Written by Richard C. Murray and E. Bodette Reeves. U.S. Geological Survey Circular 676. Washington, D.C.: USDI, 1972.

U.S. Department of Interior, Bureau of Reclamation. *West Texas and Eastern New Mexico Import Project.* Washington, D.C.: USDI, 1973.

U.S. Department of Interior. *Westwide Study Report on Critical Water Problems Facing the Eleven Western States.* Washington, D.C.: USDI, 1975.

U.S. Department of Interior, U.S. Geological Survey. *Estimated Use of Water in the United States in 1975.* Written by Richard C. Murray and E. Bodette Reeves. U.S. Geological Survey Circular 765. Washington, D.C.: USDI, 1977.

U.S. Department of Interior. California Department of Resources, and California State Water Resources Control Board. *San Joaquin Valley InterAgency Drainage Program, Final Report.* Washington, D.C.: USDI, 1979.

U.S. Water Resources Council. "Procedures for Evaluation of National Economic Development (NED) Benefits and Costs in Water Resources Planning (level C); Final Rule." *Federal Register,* vol. 44, no. 242, (1979).

U.S. Water Resources Council. *The Nation's Water Resources 1975-2000; Second National Water Assessment.* Washington, D.C.: U.S. Water Resources Council, 1978.

van Schilfgaarde, Jan. Director, U.S. Salinity Laboratory, Riverside, California. Personal communication, February 1980.

CHAPTER 5

Energy Dependence and the Future of American Agriculture

OTTO C. DOERING, III

Since the spot fuel shortages and Arab oil embargo in 1972 and 1973, a great deal of attention has been given to the role of energy in agricultural production. The many questions being raised can be grouped into three broad areas of concern:

1. What is our present pattern of production and energy use in agriculture, and how did we arrive at it?
2. What are our options for changing energy's role in agricultural production in the future?
3. What are some of the possibilities for producing energy from agriculture?

These concerns are very much interrelated; yet a great deal of discussion and writing about the future contains little background on why we are where we are today. Our ability to manipulate agricultural production systems will depend to some extent on our understanding of these systems, their limits, and potentials.

While concentrating on agricultural production, we need to recognize that production is only one portion of the entire food and fiber system. In 1974, 16.5 percent of the total energy used in the United States was related to the production, processing, marketing, and consumption of food; approximately 3 percent was for food production.* Dividing this energy input into the different components of the food system in a given year, we get the ranking shown in Table 1. One flaw of this breakdown is that the energy used for transportation is understated because it is very difficult to estimate for each segment of the food system. It is

---

*Historical data presented in this paper are all from USDA sources.

Table 1.   Relative 1974 Energy Use in the Food System

Activity	Percent of Food System Total
Consumption & Preparation	43%
In the Home — 26%	
Away from Home — 17%	
Processing	29%
Production	18%
Marketing	8%
Transportation (underestimated)	2%

SOURCE: FEA, 1976a

clear from this breakdown, however, that agricultural production is by no means the major user of energy in the food and fiber system. Given this perspective, this paper first looks at current patterns of operation and energy use in agricultural production. Some options for changing energy's role in agricultural production then are considered. Finally, some of the alternatives being discussed for producing energy from agriculture are examined.

WHERE WE ARE TODAY

Energy is only one of the many resources used in agricultural production. Recently we have tended to focus on energy almost to the exclusion of other resources. In doing so, we sometimes lose our perspective on the total system, a perspective that is essential in understanding shifts in the use of different resources. To regain perspective, we need to look at past trends and attempt to understand some of the reasons for the resource shifts that have taken place.

The American crop production system is usually characterized as being highly energy intensive. It is also highly productive. Table 2 illustrates this productivity, the land resource use, and the disposition of the product.

The U.S. population has increased since 1950, but the level of population increase has been more than matched by increases in the level of crop production. Crop production per acre increased slightly faster than overall crop production, reflecting the change in cropland used for crops. Cropland used for crops declined some in the 1950s and 1960s, particularly as government programs took land out of production. Currently cropland is at almost the

Table 2.   Index of U.S. Population and Crop Output (1950 = 100)

Year	U.S. Population	Crop Production	Crop Production per Acre	Cropland Used for Crops	Export Acres
1950	100	100	100	100	100
1960	119	121	128	94	128
1970	134	132	148	88	144
1975	140	158	163	97	200
1978	143	171	176	97	226

SOURCE: USDA, ESCS, 1980

same level it was in 1950. Export acres have increased dramatically, as crop production continued to outstrip the rate of domestic population growth.

Note that the acres in production in 1950 are not necessarily the acres in production in 1978. Some high-quality agricultural land has been taken for other uses and may have been traded for marginal land. At the same time, marginal land may have been improved to the point at which it is now highly productive. Thus, there have been changes in land quality and location that reflect a continuing dynamic pattern of adjustment to changing demands for land. These changes, in turn, have influenced the quantity and quality of other resources used in agricultural production.

Production assets per farm and per farm worker have increased greatly over the last four decades. Production assets per farm worker are greater than those for the industrial worker. Part of the increase represents the addition of capital stock—machinery, land, and so forth—or improvements (such as drainage structures) on a per farm or per worker basis. However, some of the increase, especially since the early 1970s, represents an inflation-related increase in the price of land.

Production assets are becoming increasingly concentrated, partially reflecting a concentration of the primary asset for agriculture—namely, land. As Table 3 illustrates, the number of farms has decreased dramatically, while the size of farms has increased. Within the total population of farms, a decreasing number of large farms is supplying an increasing proportion of the total production. In 1960, 21 percent of the farms produced 73 percent of the total value of farm output. In 1976 (with 30 percent fewer farms), 17 percent of the farms produced 90 percent of the total value of farm production. Concurrent with this trend is the

Table 3.   Inputs for Agricultural Production

Land

Year	Number of Farms	Land in Farms Number of Acres (1,000)	Acres per Farm
1940	6,096,799	1,060,852	174
1950	5,382,162	1,158,566	215
1960	3,962,520	1,175,646	297
1970	2,954,200	1,102,769	378
1975	2,803,480	1,086,025	387
1979	2,330,000	1,048,768	450

Production Assets

Year	Per Farm	Per Farm Worker
1940	$ 6,200	3,300
1950	17,200	9,400
1960	42,100	21,100
1970	86,904	55,822
1975	153,723	96,562
1979	258,824	172,637

Farm Work Force

Year	Total Workers (thousands)	Workers per Farm
1940	10,979	1.80
1950	9,926	1.84
1960	7,057	1.78
1970	4,523	1.53
1975	4,342	1.55
1979	3,937	1.69

SOURCE: USDA, 1978

decline of farm work; this decline in labor resources committed to agriculture is a mirror image of the increasing commitment of some other resources.

Table 4 lists the petrochemical input in the form of fertilizer and various pesticides. Total chemical nutrient application has been increasing at a rather steady rate over the last four decades. The use of herbicides on many crops has increased a great deal, while the use of insecticides for food and feed grains has increased slightly.

Table 4.   Use of Agricultural Chemicals

Fertilizer

Year	Tons (1,000)
1941	9,296
1951-55 average	22,467
1961-65 average	28,607
1971-75 average	41,438

Pesticides

	Farm Use of Herbicides by Crop (in million pounds)		
	1966	1971	1976
Corn	46.0	101.1	207.1
Soybeans	10.4	36.5	81.1
Wheat	8.2	11.6	21.9
Cotton	—	19.6	18.3
Sorghum	4.0	11.5	15.7
Rice	2.8	8.0	8.5

	Farm Use of Insecticides by Crop (in million pounds)		
	1966	1971	1976
Cotton	—	73.4	64.1
Corn	23.6	25.5	32.0
Soybeans	3.2	5.6	7.9
Wheat	—	1.7	7.2
Alfalfa & Other Hay	—	2.5	6.4
Sorghum	—	5.7	4.6

SOURCE: USDA, ESCS, 1978

The trend toward increased use of pesticides is further illustrated by Table 5, which gives the acres treated for a given crop in
a given year.

Table 6 shows the trend in tractors and horsepower and also
shows the decline in person-hours required per 100 bushels of
corn for grain. There is an important qualitative distinction to be
made in looking at the trend in machinery. While the number of
tractors actually began to decline slowly after 1966, the horsepower continued to increase. Thus, the capacity to do work with
machinery was still increasing.

Summaries of trends can hide important complexities. For

Table 5.   Acres Treated with Pesticides

Herbicides:
Percentage of crop acres treated for selected crops and years

| | Percentage of Acres Treated | | | | |
Crop	1952	1958	1966	1971	1976
Corn	11%	27%	57%	79%	90%
Small Grain	12%	20%	29%	36%	37%
Cotton	5%	7%	52%	82%	84%
Pasture and Rangeland	—	1%	1%	1%	2%

Insecticides:
Percentage of crop acres treated for selected crops and years

Crop	1952	1958	1966	1971	1976
Corn	1%	6%	33%	35%	38%
Cotton	48%	66%	54%	61%	60%

SOURCE: USDA, ESCS, 1978

example, it can be summarized that real estate, as a farm input, remained relatively stable from 1960 to 1978. But one question that would need to be asked would be: What is the nature of the real estate or land input? It appears to have been relatively stable, but what has the change in location or quality meant in terms of the requirements for other inputs? We do know that the increase in the proportion of irrigated acres has increased the energy intensity of agriculture generally, and fertilizer has been an important part of the growth in agricultural chemical use. In fact, fertilizer is by far the largest energy input, as Table 7 illustrates.

The figures in Table 7, however, are "average" figures and really represent no single type of crop enterprise. As will be shown later in this paper, energy for irrigation may be the largest single energy input on an irrigated farm, and pesticides can be one of the largest energy inputs for crops like cotton.

SOME REASONS FOR CURRENT ENERGY-USE PATTERNS

There are a number of reasons why we are where we are today in our patterns of resource use. These reasons, which are economic, institutional, biological, and technical, are almost always not mutu-

Table 6.   Tractors, Horsepower, and Labor Requirements

Tractors and Horsepower

Year	Number of Tractors (x10³)	Total Horsepower (x10⁶)
1941	1,665	45
1951	3,678	101
1961	4,695	158
1966	4,783	182
1971	4,584	206
1976	4,434	228
1978	4,370	238

Person-Hours per 100 Bushels of Corn for Grain

Year	Person-Hours
1935-39	79
1945-49	53
1950-54	34
1955-59	20
1960-64	11
1965-69	7
1974-78	4

SOURCE: USDA, ESCS, 1980

Table 7.   Direct Energy Use in Farm Production (1974)

Input	Trillion Btu	Percent
Fertilizer	621	31%
Irrigation	261	13%
Farm Vehicles	261	13%
Preharvest Field Operations	249	12%
Harvest Operations	201	10%
Grain & Feed Handling & Drying	157	8%
Livestock Care	121	6%
Pesticides	95	5%
Other	48	2%

SOURCE: FEA, 1976a

Table 8.   Prices Paid for Land, Energy, Labor, and Fertilizer

Index: 1950 = 100

Year	Average Value of Farmland	Cost of Gasoline	Farm Wages	Cost of Fertilizer
1950	100	100	100	100
1951	115	101	110	106
1952	126	102	118	108
1953	128	105	121	109
1954	126	107	120	110
1955	131	112	121	108
1956	139	115	126	106
1957	150	120	131	106
1958	158	118	135	106
1959	171	119	144	106
1960	179	122	148	106
1961	182	121	151	107
1962	191	121	155	106
1963	199	121	159	106
1964	212	119	163	105
1965	224	121	171	106
1966	243	123	185	103
1967	259	127	199	102
1968	276	128	216	96
1969	289	132	238	89
1970	300	134	255	90
1971	313	136	268	93
1972	337	136	284	96
1973	380	149	309	104
1974	477	203	354	171
1975	545	213	383	222
1976	600	224	419	189
1977	702	235	451	185
1978	751	250	483	184
Total Increases	651% (Land)	150% (Energy)	383% (Labor)	84% (Fertilizer)

SOURCE: Adapted from USDA, 1955-1979

ally exclusive. An economist might identify the driving force as economics, but many other factors have to be involved as well.

Table 8 summarizes some of the economic reasons for our current production practices and patterns of resource use. Between 1950 and 1978, land and labor became increasingly more expensive in relative terms than energy and fertilizer. Actually, this trend holds even from a 1910-1914 base period. From 1950 to 1978, energy prices went up roughly twice as much as fertilizer prices, but labor prices went up more than twice as much as energy prices, and land prices increased more than four times as much as energy prices.

There has been an increasing incentive over the long term to use low-cost resources to improve the productivity of high-cost resources or substitute for them. And such incentives have led to the adoption of new technology. Institutions have played an important role in this endeavor. The publicly funded research and extension institutions often were critical in the development and practical application of new technology. Given the highly dispersed and competitive nature of the agricultural production sector, a publicly funded research and extension system has been responsible for developing technology that might not have been developed by the private sector because of the difficulty the private sector has in capturing the benefits of investment in research. In addition, other government institutions and programs provided credit and reduced the risk of commodity price fluctuations, both of which improved the environment for resource shifts and the adoption of new technology.

The influence of government policy on the cost of energy has also been important. Unlike most other countries around the world, we do not have a long tradition of using taxes on petroleum as a major revenue source. Petroleum fuels were available in large quantities at relatively low prices. We also have had, and continue to have, a policy of keeping the price of natural gas low by direct government price regulation. This has certainly been a factor in the relatively low price of fertilizer, although the capacity of the industry appears to have a greater impact on the final price than does the cost of natural gas feedstocks. Finally, electricity was provided to farms through the Rural Electrification Act on a more favorable basis than might have been the case without such a government program.

To some extent, the agricultural programs of the 1950s and 1960s were both the product of a shift in resource use as well as a further stimulus to such shifts. The basic agricultural problem of the 1950s and 1960s was that we had improved our productivity to the point at which we could neither use all the agricultural products we produced nor sell them to the rest of the world. To support farm incomes through increased agricultural prices, the government had to require the idling of a certain amount of farm-land so that the Treasury costs of such a program were kept within reason. This, in turn, gave a great incentive to farmers to increase the productivity of the land remaining in production so that they could reap the largest benefit possible from the guaranteed government prices. Farmers tended to keep their best land in production, the land that yields the greatest return to the increased addition of inputs like fertilizer. Increased returns then were capitalized into the value of land and were a factor in its price increases. The net effect of government activities was to hold down the relative cost of energy and chemicals while increasing the cost of land and labor (considering the direct and indirect effects of minimum-wage laws).

It is difficult, however, to isolate the influence of economic forces, government programs, and institutions. Was the increasing intensity of energy use in agricultural production a cause of the concentration of U.S. farm production in larger units, a result of it, or just a benign factor that exercised little influence in one direction or another? Was the urge to save labor critical for the adoption of the tractor? Or was it the fact that the tractor released areas of high-quality pasture land for cultivation, while at the same time releasing grain grown for horses to other uses, thus increasing both the cultivated area of a given farm and the proportion of the product that could be marketed off the farm for cash? Given the complexity of the system, it becomes almost impossible to assign simple deterministic causality.

Other influences notwithstanding, we can say that increasing amounts of energy have been used in agricultural production for the following basic reasons:
1. To reduce the use of human and animal labor;
2. To increase production per unit of land, labor, or other limiting input;
3. To reduce risk of crop failure or spoilage during growing, harvesting, and primary processing operations.
This categorization provides an alternative interpretation of past trends, based on biological and technical factors as well as the drive to reduce risks.

The continuous increase of tractor horsepower on farms has been a major means for reducing the use of human and animal labor. Farm electrification was a parallel development that allowed a compact and relatively inexpensive application of energy to all farmstead tasks, some of which were not even attempted on the farm previously.

The increase in fertilizer use, the increased use of other petrochemicals for weed and insect control, and the use of natural gas, diesel fuel, or electricity for irrigation are good examples of applying energy for increased production. All have tended to enhance greatly production per unit of land and per unit of other nonenergy inputs.

Some of the inputs mentioned also are used to reduce risk. This is certainly true of the preventive use of insecticides, and of the use of herbicides in combination with conventional tillage and the use of energy for supplemental irrigation. There is a link in many cases between the enhancement of production and the reduction of variability due to natural forces like weather and pests. A similar combined effect occurs in attempts to reduce risk in harvesting grains. Ever-larger equipment with greater capacity allows more timely harvesting and increases in yield. The improvement in timeliness also may allow fall tillage, which can result in timely spring planting and enhanced yield the following year.

The Interrelated Nature of Agricultural Production

While the above categorization of energy use provides a functional understanding of the reasons for the application of energy to agricultural production, we still need to recognize how the total system (both biological and technical) constrains the use of energy and how that energy may be modified.

Table 9 should be useful in understanding some of the interrelationships in crop production. On the left side are some basic natural factors determining yield. Over time, people have attempted to modify one or more of these factors to improve yields and/or to reduce yield variability or risk. The control factors are listed on the right side of the figure. As the control factors are applied, the use of energy is increased.

There are also important second-order interrelationships. As an example, once irrigation has been adopted to modify the weather and improve crop yield, then earlier practices of fertilization or pest control, which might have been appropriate to nonirrigated situations, have to be redefined and adjusted. Thus, not only do the control factors interact with the basic factors they modify, they also interact among themselves and with other basic factors.

Table 9.   Yield Factors and Control Factors

Basic factors determining the level of yields	Associated control factors which may be brought to bear through management as influenced by size, etc.
Weather _____	Irrigation, drainage, frost control, crop drying
Soil quality _____	Cultivation practices, terracing, crop rotations
Genetics_____	Choice of variety
Plant nutrition _____	Fertilization, crop rotations
Timing _____	Labor availability, machinery stock
Weeds_____	Crop rotations, herbicides, cultivation practices
Pests _____	Choice of variety, crop rotation, pesticides, cultivation practices

SOURCE: Doering, 1977

The degree to which the interaction of control factors and basic yield factors results in higher yields is subject to a number of "quality of management" factors which may be correlated one way or another with size and/or other attributes of the physical/ human farm environment.

The more severe the control factor, the more impact the second-order interactions are likely to have.

These multiple interactions demonstrate the necessity for a systems approach in analyzing the impact of specific energy inputs in agriculture. Figure 1 gives an accurate guide to understanding a given cropping system, assuming that a change in one component is likely to have numerous direct and indirect impacts on a number of other components in the system. In this case, cropping systems are seen as being bound by weather and subject to such control factors as management, genetics, and cropping patterns.

One result of the multiple interactions is the difficulty of predicting the outcome of any movement away from an existing favorable position that involves an intensive application of control factors. A large number of new multiple adjustments would be necessary and critical in shaping the new final input/output relationship. This is an important factor in assessing the degree of flexibility in moving from one cropping system to another so that the energy requirement can be changed.

System Considerations

There are aspects of current production systems that can be explained in terms of responses to system considerations as well as they can be explained in terms of economic or agronomic concerns. The field shelling of corn is a good example.

By the late 1960s, the majority of the corn produced in the Corn Belt was field shelled with combines and artificially dried. Only a decade earlier, this corn had been harvested with pickers and handled in ears. Ear corn is roughly twice the volume of shelled corn for an equal amount of grain yield. This, however, does not necessarily slow the harvesting machine. A six-row picker can move through the field and perform its task as quickly as a six-row picker-sheller. The potential bottleneck is in handling the ear corn while getting it out of the field and storing it. Shelled corn has a tremendous advantage because it is free flowing, which allows it to be moved by an auger or by gravity. The high yields of new hybrid corn varieties severely taxed ear-corn handling systems to the point at which the handling bottleneck and high labor requirement, along with some advantages to shelled corn per se, induced farmers to adopt shelled corn systems.

There is also no question that shelled corn systems will give farmers greatly increased marketing flexibility. If the shelled corn to be sold as grain is first handled as ear corn, it is not dry enough to enter the cash grain market until late winter, when enough natural drying has taken place to reduce its moisture content to 15 percent or below. As long as the bulk of the corn was used on the farm or used locally, this did not pose any particular problem. However, with the growth of the export market, the value of flexibility in the timing of grain marketing became sufficient to further induce a movement to shelled corn systems, since these provided a marketable and more easily storable product almost immediately after harvest.

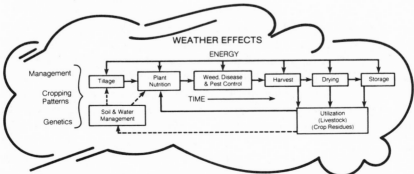

Figure 1.   The Dynamic Effects of Energy, Weather, Management, Cropping Patterns, and Genetics on the Crop-Production System
SOURCE: Doering, 1977 .

While marketing flexibility or reduced labor requirements can be figured easily on a benefit/cost basis, many characteristics of the ear-corn/shelled-corn systems may not have been costed directly. These characteristics entered farm decisions in terms of such things as the necessity for timely harvesting when the corn crop reached maturity (in spite of high moisture) to prevent field losses and to allow fall plowing. The trade-off between field losses and natural drydown is such that harvesting at relatively high moisture levels, thereby avoiding field losses, remains profitable until drying fuel costs are two or three times the current price. Fall plowing also has become a necessity in many operations to ensure timely planting in the spring. A combination of all these factors induced farmers to move to shelled corn systems.

The Nature of Agricultural Energy Demands

Agricultural energy demands are quite different from those of most manufacturing processes. Once set in motion, the agricultural process is virtually irreversible, and a partially produced crop is not salvageable. With corn production, more than half the total energy inputs have been committed once the crop has been planted. Something like a drought may result in little or no output, and there is no way that the substantial energy input already committed can be retrieved.

The same factor operates with even greater potential loss once the product is produced. If harvesting and processing cannot be accomplished in a timely manner, the biological process continues and takes its toll on the final product.

The weather-dependent nature of agriculture makes the absolute level of energy requirements highly unpredictable from one year to the next. Of the three uses of energy in agriculture—to reduce human and animal labor, to raise productivity, and to reduce risk—energy used to raise productivity is probably the most stable in terms of farm demand from one year to the next. However, the energy used to reduce risk, such as that for pest control, crop drying, and irrigation, is highly variable, depending on weather and a host of other factors beyond the control of the agricultural producer. The energy used for cultural practices (such as tillage or fertilization) is also highly variable in its tie with the energy used to reduce risk, variability, and spoilage. When weather makes the application of herbicide ineffective, either mechanical cultivation or respraying, with its energy input of both petrochemicals and tractor fuel, is necessary. A wet fall may mean greater fuel requirements for harvesting corn and the need for two or three times as much drying fuel as is needed in a very dry year. Excellent examples of such contrasting fuel demands for the same crops in the same regions have occurred over the last three to five years.

If energy inputs are to enhance the biological production process, they must be applied when they can have the most effect on that process. There is little use applying fertilizer after the plant has already passed through several of its important growth stages. Plowing and planting must be accomplished early in the spring months so that the plants have as much leaf area as possible when peak solar radiation is available in early summer. The agricultural producers in a region need their fertilizer, diesel fuel, and other inputs at specific times, coinciding with favorable weather conditions and the biologically based plant requirements. In the Corn Belt, a great deal of energy is needed in a short period in the spring for fertilization, plowing, and planting. Another peak energy demand occurs in the fall for harvesting and grain drying. It would be impossible to spread these energy needs out over the year without greatly reducing agricultural production and the efficiency of energy use in agriculture.

The specific contribution of a single energy input in the production system is often almost impossible to isolate or measure accurately. When there are many energy inputs to the production process, something of a synergistic process occurs. This makes it difficult to assess in a useful way the contribution of only one of the inputs. An example of this is the attempt to isolate the effect of herbicide use in the intensive production of row-crops. By withdrawing herbicide use from such a system, yields often will be severely reduced since weed populations also benefit from the high fertilization rate and other stimulative aspects of the intensive system. Yet, to assume that the yield difference is due to herbicides alone is misleading. The high crop yield with herbicide depends on factors such as high levels of fertilization and adequate moisture availability as much as on herbicide use. In other words, we are dealing with joint effects. Insofar as this synergism exists, attempts to assess the contribution of herbicide energy or fertilizer energy on a single input basis can be misleading. It becomes especially misleading in energy input/output analysis, which views inputs as separable and distinct.

The interrelated nature of agricultural production makes its energy problems extremely complicated; biological constraints make these problems even more difficult. An effort to save energy by withdrawing it from one part of the system, for example, may be counterproductive on the basis of overall resource use and productivity in the system. Savings of energy per se sometimes result in lower productivity per unit of energy input. There are instances when the addition of more energy to the total system would result in proportionally greater productivity and, in fact, increase output per unit of energy input.

OPTIONS FOR CHANGING ENERGY'S ROLE IN AGRICULTURAL PRODUCTION

Although we may anticipate overall energy shortages in coming years, we do not now have the information that would allow us to make intelligent choices among the alternatives that might help agriculture meet such shortages. It is not enough to know that we will have energy constraints relative to other constraints. In the past several decades, the costs of a wide range of energy forms have been so low, and the complete logistical system for assuring delivery at almost any time and place has operated so smoothly, that we have not had to consider these matters as serious constraints in the design and operation of agricultural production systems. The most convenient forms of energy were readily available and at prices that encouraged their use.

The time constraint in agriculture tends to intensify the energy-form constraint. Much of the direct energy use in agriculture is tied to a specific fuel form because of logistical requirements or type of process. The need for motive power in field operations, for instance, sets a very inflexible liquid-fuel requirement, just as high-speed processes for drying grain require a clean gas fuel with combustion products that can pass through the grain without causing contamination. Much of the timing of the production process is set by the weather and by the biological development of the product itself. If there were a shortage of the kind of fuel needed, there would be little time to convert equipment to allow the use of a different form of energy.

One of the trade-offs made to achieve higher yields through specialization and intensification of crop production is a loss of flexibility. When there was a shortage of liquefied petroleum gas (LP-gas) for corn drying in the fall of 1972 and 1973, there was virtually no other way to handle the harvest because the harvesting and storage system was locked into a specific drying system. The suggestion to return to ear harvesting was ludicrous to farmers; they had neither the ear-corn harvesting equipment nor the corn-crib capacity to handle even a small fraction of the current crop volume. Specialization, in effect, has further locked in the time, form, and place requirements for energy inputs into agricultural production.

The first energy shortages in the 1970s were not really a result of energy unavailability. The shortages of LP-gas were more the result of transportation bottlenecks and price controls than they were the result of absolute shortages. The subsequent federal propane regulations broke the market logjam caused by the price freeze, but supplies were still tight because of transportation bottlenecks. Transportation bottlenecks were also a problem in the winter of 1976/77 for both propane and natural gas.

Before defining options for changing energy's role in agricultural production, it is necessary to identify the nature of the energy shortages or relative price increases likely to occur. The most likely candidates for shortages are liquid fuels. On a short-run basis, there are almost no fuels or technologies that could compensate for a cutoff of liquid fuels. However, projections that have short-term shortages of liquid fuels resulting in diminished production proportional to the fuel shortage on a production function basis are not very realistic. With stringent conservation practices, there is probably enough flexibility to absorb a moderate reduction in liquid-fuel availability without a proportional production loss. In practice, it looks as though federal policy would give priority to liquid fuels to maintain essential cropping practices in a time of emergency. The immediate shutoff of liquid fuels, therefore, is probably not a realistic concern.

The important question for adjustments in liquid-fuel use involves long-term usage. An adjustment in cropping systems and cultural practices could be precipitated by the price of liquid fuels and by the threat of absolute shortages. One alternative often mentioned is no-till or minimum-till systems for major grains. Several factors need to be kept in mind in assessing this alternative, however. There is already a great deal of minimum or no-till production, and not all types of land are suitable for such systems. Thus, the potential adjustment is limited to that land that is suitable for these systems but not yet under them. Furthermore, the energy savings with no-till or low-till systems are less than they might appear to be because of the petrochemical energy necessary to produce the extra herbicides and pesticides such systems require and because many of the high draft soils in which tillage fuel savings would be greatest are not very adaptable to low or no-till systems. In corn production, we expect only one gallon of gasoline an acre as an across-the-board savings for changing to no-till or minimum tillage, even though specific case savings may be higher.

In general terms, especially after comparing the relative trends of energy costs and labor costs (Table 8), it is difficult to imagine price changes bringing about a substitution of labor for energy in those cases when energy was applied in agriculture to reduce human and animal labor. As an example, the conversion to no-till systems can, in some cases, reduce labor requirements; in fact, the reduced labor requirement may have been one reason for the adoption of the no-till system. In this case, there is almost a complementary relationship between energy and labor.

We do not expect shortages of electricity for general farm use where it has replaced human and animal labor. There also appears to be little, if any, cost incentive to move these farmstead operations, such as feed grinding, back to human or animal labor or to relocate them off-farm, where other power sources can be used on a centralized basis. Some of the centralized activities might well involve more transportation energy than is currently used in the on-farm activity, and that would require scarce liquid fuels, whereas much of the electricity can be generated from coal.

In most agricultural production, the energy that has been added to increase yields is the largest energy use in the total process. In Table 10, intensive conventional corn production in the Corn Belt is used to compare the magnitude of energy inputs in gasoline fuel equivalents. It makes sense to reduce energy in those parts of the production system that are comparatively large users of energy. We can achieve more by saving 10 percent of 40 gallons than we can by saving even 50 percent of 5 gallons. This is one reason why the energy used in fertilization has been of special concern.

Table 10.   Energy Inputs for Corn Production

Input	Per Acre Energy Content in Gasoline Equivalent
Fertilizer (Mainly Nitrogen)	40 Gallons
LP-gas (Crop Drying)	20 Gallons
Tillage and Cultivation	10 Gallons
Herbicides and Insecticides	5 Gallons

SOURCE: Doering and Peart, 1977

Much publicity has been given to cropping systems based on rotations that provide nitrogen for fertilization through the use of legumes. Such cropping systems can save considerable amounts of energy through nitrogen replacement. A number of these cropping systems were compared by simulation with conventional systems, and the results are given in Tables 11 and 12. These systems were simulated for a farm with 600 acres of top-quality Corn Belt land. It was assumed that geneticists had solved the cold tolerance problem and one of the legumes, winter vetch, would perform quite well under colder conditions than it normally has done in the past. In physical output/input terms for both energy and protein, the legume systems did perform better than the more traditional continuous corn, corn/soybean, or corn/soybean/wheat systems. However, for the given acreage, the continuous corn system produced a much greater amount of digestible energy in absolute terms.

The economic performance of these systems depends on the relative prices of the cash crops, especially corn and soybeans. Alfalfa is a cash crop as well as a producer of nitrogen; the problem is that the alfalfa market is relatively limited, and the economic feasibility of using alfalfa to produce nitrogen depends to a large extent on the price the farmer might be able to receive for alfalfa. If a large number of farmers began growing alfalfa, there would be a real drop in the price the farmer would receive for this crop. The vetch has no crop value, and the system's poor economics reflect this. There are a few grain farmers adopting legume rotations to obtain nitrogen. There are also farmers who have mixed crop/livestock enterprises, using legume rotations and animal wastes for fertilizer. Such farms usually include pasture as

Table 11. Summary of Six Cropping System Simulations, Annual Averages, 600 Acres of Class I Well-drained Level Land

System:	A[b]	B[c]	C[c]	D[e]	E[g]	F[i]
Rotation:[a]	Cont. Corn	C-C-S	C-C-S-W/S	3C-3S-W/3A	C-S/V	C-S/V Min. Till.
Acres						
Corn	600	400	300	200	300	300
Soybeans	0	200	300[d]	200	300	300
Wheat	0	0	150	66.7	0	0
Alfalfa	0	0	0	200[f]	0	0
Wheat-Vetch	0	0	0	0	300[h]	300[h]
Production, 10⁵ pounds						
Corn	46.6	31.6	23.5	15.9	22.8	23.4
Soybeans	0	5.7	7.1	5.7	8.4	8.5
Wheat	0	0	4.9	2.2	0	0
Alfalfa	0	0	0	13.1	0	0
Total Input, 10⁹ Btu						
Energy Equivalent	4.98	3.68	3.42	2.78	2.55	2.62
Total Output Digestible						
Energy, 10⁹ Btu	30.3	24.7	23.5	22.9	20.9	21.4
Protein, 10⁵ pounds	3.31	4.02	4.35	5.11	4.22	4.30
Output/ Energy Input						
Dig. Energy, Btu/Btu	6.08	6.71	6.87	8.25	8.20	8.18
Dig. Protein, pounds/10⁵ Btu	6.65	10.92	12.74	18.37	16.54	16.44

SOURCE: Doering, 1977

a. (C = corn, S = soybeans, W = wheat, A = alfalfa, V = winter vetch and wheat)
b. Timeliness is excellent.
c. Timeliness is excellent. Machinery is much larger than needed.
d. 150 acres of this is double-crop soybeans.
e. Timeliness is good except wheat harvest delays second cutting of alfalfa and fourth cutting interferes with corn harvest.
f. 66.7 acres of this is interseeded in standing wheat.
g. Corn and soybean plantings are late because cover crop must be plowed down.
h. Wheat-vetch is intercropped between soybeans and corn as a winter cover crop.
i. Minimum tillage option. Corn is coulter-planted and soybeans are drilled in 7" rows. There is no cultivation. Timeliness is good except corn and soybean plantings are slightly late.

Table 12. Net Returns to Management from the Six Cropping Systems

System:			A	B	C	D1[a]	D2[a]	E	F
			Cont. Corn	C-C-S	C-C-S-W/3A	3C-3S-W/3A		C-S/V	C-S/V Min. Till.
Prices, $/bu.									
Corn	Soybeans	Wheat							
Present Plan									
$2.36	$5.17	$2.86	$43,500	$41,500	$37,000	$40,500	$59,000	$31,500	$28,000
$2.36	$5.17	$2.86	$42,500	$46,500	$43,500	$45,500	$64,000	$39,000	$36,000
$2.36	$6.56	$2.86	$43,500	$55,000	$53,500	$53,500	$72,500	$51,000	$48,000

SOURCE: Doering, 1977

a. The two D systems include alfalfa yielding six tons/A, with system D1 having a hay price of $60 per ton and system D2 having a hay price of $80 per ton.

well as cropland to allow an economical combination of livestock and grain enterprises in terms of possible rotations, the use of forages, and the utilization of manure.

Natural gas is the major feedstock for the production of nitrogen in the United States. While we may doubt the existence of a natural gas glut, there is good evidence that chemical nitrogen will still supply most of our cash-crop nitrogen requirements for the next 20 or 30 years. There are three reasons for this. First, if large numbers of cash-grain farmers adopt legume rotations, they will forfeit a portion of their land to legume rotation, which will probably provide less cash-crop value than the feed or food grain grown previously. Second, projections of chemical nitrogen fertilizer prices for the next decade or so are not out of line with projections for crop prices or the costs of other inputs in relative terms. Finally, even if we insist on short-sighted natural gas policies that result in shortages and high prices, there are other feedstocks that can be used for nitrogen. Although the cost of these feedstocks will be higher, it will probably not be high enough in relative terms to encourage widespread substitution by legumes. On the other hand, genetic improvement in the ability of legumes to fix nitrogen or improvement in the adaptability of legumes to tightly scheduled cropping systems might well result in the increased use of legumes for nitrogen production.

When irrigation is used to increase production, particularly in dry areas where it supplies most of the water required, irrigation often becomes the largest energy input. Whether the fuel is natural gas, diesel, gasoline, or LP-gas, the energy input is so large that the cost of this input strongly influences the production system. An illustration from the high plains of Oklahoma is given in Table 13. It indicates the relative magnitude of energy requirements in

Table 13. Major Energy Inputs—Irrigated Corn in Oklahoma High Plains

Per Acre	
Input	Btu Energy Content (x 10³)
Fertilizer: Nitrogen (200 lb./acre)	6,220
Tractor Fuel: Diesel (13.4 gal./acre)	1,876
Irrigation: Natural Gas (14.6 MCF/acre)	15,330

SOURCE: Eidman, 1975

conventional corn production when water must be pumped for irrigation from depths of 350 to 400 feet.

There are several types of irrigation systems that do not have such a high-energy budget—for example, gravity systems, which have little or no direct energy requirement, and surface water that is pumped for spraying or distribution. There are also states such as Washington that have had abundant electricity from the major federal hydroelectric projects. If their electric rates remain at the current low relative costs, then pump irrigation remains economically feasible.

One immediate impact from higher energy prices will be the elimination of irrigation, for all but the highest-value crops, in areas of deep aquifers. As an example, it has been estimated that by 1995 increasing natural gas prices will result in the termination of the irrigation of crops such as cotton and sorghum in the Texas High Plains (Young and Coomer, 1980).

On-farm crop drying is an energy use that is likely to be modified—but neither to the extent nor as soon as is projected for pump irrigation. Most on-farm grain drying uses LP-gas in systems in which the heat and products of combustion are vented through the grain. These systems do not allow the use of liquid or solid fuels. The only good substitute for LP-gas in these systems is natural gas; where pipelines exist in rural areas, natural gas is already used because of its relatively low, government-controlled price.

Two alternatives to LP-gas that have been under development appear to be technically and economically feasible, given current LP-gas prices, which range from 50 to 60 cents per gallon. One of these is gasifying corn cobs and using the low Btu producer gas to operate current drying equipment. The other is using electric heat pumps for low-speed drying. This system is economic in comparison with electric resistance drying systems.

Given that 2/3 of our LP-gas comes from natural gas, and this does not appear to be in short supply over the near future, there is not much incentive to change grain drying systems on the basis of threatened absolute shortages. Since grain drying is still a relatively modest portion of total costs, these costs do not exert much force in changing drying systems at a greater rate than normal equipment replacement. Initially, because new systems have only a slight cost advantage, conversion will only occur if these systems are seen as equally reliable in preserving the crop, which represents not only the sunk costs, but also the potential year's income. Averting risk is particularly important at this point in the production process.

The Effects of Changing Energy Prices

There are any number of questions being raised about the effects of shifts in energy prices and/or availability on such things as the type of agricultural production, quantity of production, location of production, and level of commodity prices. Answers to these questions depend on the kind of environment agricultural production will have over the next several decades. With the sort of energy environment that has been described in this paper, most of the effects would be very modest.

Level of Commodity Prices. Given such things as government guarantees of liquid-fuel availability and the continued availability of natural gas (with phased decontrol), there does not appear to be an energy factor that would severely disrupt the level of production. Levels of commodity prices are more likely to continue to be determined by such factors as the strength of export markets, changes in technology, and variations in weather, rather than by changes envisioned in the price and availability of energy over the next several decades.

Type and Quantity of Production. Some concern exists about the continuation of intensive production and the mix of commodities that will be produced. Extreme low-energy scenarios picture farm production units that are largely self-contained in supplying much of their energy demand. Under this scenario, the mix of commodities available changes, and diets are made more efficient in terms of energy conversion. That is, instead of humans consuming cattle, which have been fed on corn, more high protein vegetables are consumed directly by humans. The energy efficiency of our diets, therefore, is increased, and a lower level of feed grain production does not affect the nation adversely. One difficulty would be supplying the form of energy needed on the farm solely from farm sources. It does not appear that energy prices or lack of availability will drive us very far in this direction over the next several decades.

Transportation. It appears that, gradually, the location of some farm activities will shift. For example, it has become more profitable recently to feed Kentucky calves at home with Kentucky corn, rather than ship them to Texas feedlots, which is the current

practice. Much of this change in profitability (about $5 per hun-
dredweight) is due to increased transportation costs. Yet, there
will not be an immediate shift in the location of feeding for several
reasons. First, it will take time to depreciate fully the cost of
feeding facilities in Texas. Sunk costs will exert a strong influence
on maintaining current systems, and owners of existing facilities
will take a reduction in profits to stay in business. It also will take
time to confirm that high transportation costs are long term,
thereby reducing the risk of investing in new, closer facilities.
These locational shifts will likely take place eventually, however,
and they will change the location of numerous agricultural activities and will reduce the geographical concentration of some specialized processing and growing activities.

Some changes also will occur in the comparative cost of different commodities—or in the comparative cost of the same commodity in different forms. A good example of the latter is dehydrated potatoes. A fresh Idaho potato requires less energy to
get to the final consumer's plate than does an instant or dehydrated potato only if it does not have to travel too far from Idaho.
The transportation cost for a fresh potato is so much greater than
for a dehydrated potato that, in savings on shipping costs, the
dehydrated potato quickly overcomes the energy required for
dehydration and processing.

The Total Food System. While this paper focuses on agricultural production, the total food system is the actual context for
changes and adjustments. To the extent that energy prices and/or
availability modify other parts of the food system, changes may be
forced on the production sector. It is almost impossible to specu-
late on such changes. Since intuitive comments like "there will be
less energy-intensive food processing" contradict a number of
specific cases like the dehydrated potato, generalizations can be
erroneous. Factors such as consumer tastes, shelf life, and the
energy requirement of maintaining already processed food will be
more important to a processing decision than the energy used in
the actual processing.

Technology

Technology already is being developed and applied to allow us to
continue current energy-use patterns, at least to the point at
which we do not have to suffer a severe write-down of the capital
and management investment in current technologies and practi-

ces. For example, tremendous advances are being made in irrigation scheduling and other techniques to minimize the use of water and energy in irrigation. This will extend the life of some irrigation systems and will expand the inventory of crops to be irrigated, though overall irrigation will decrease. As another example, integrated pest management systems are becoming more cost effective and are receiving more interest and investment in development because of the increasing cost of petrochemicals. This is in addition to the initial impetus, environmental restrictions. Plant protection is being accomplished with a different mix of inputs.

One interesting development is the expanding role of information, be it related to irrigation scheduling or pest management. There is an increasing ability to gather and interpret information that allows us to use resources more efficiently. We are beginning to find that there is a rate of substitution between information and energy inputs into production. High energy costs have greatly increased the value of information and its application through skilled management. While many talk of labor substituting for energy, there is not likely to be a return to manual farm practices. What is more likely to occur is farmers collecting more detailed crop information in the field and adjusting their management and practices accordingly. One inhibition to the development of information systems is that benefits may not be captured by the private sector—which reduces private-sector incentive to develop them. The public sector research establishment may have to be given the resources to take a leading role here.

Institutions and Change

It is no more than a truism to state that institutions will influence the course of adjustments and changes. However, there are a number of important issues yet to be decided that will set a distinct tone to decisions made by institutions. One example is the choice of who makes decisions and another is the balance between equity and efficiency criteria in energy pricing.

The choice of who makes decisions is especially critical for agriculture. There is a belief held by many that energy problems are subject to uniform technical solutions. Some of this is reflected in the general public's willingness to spend lavishly on energy research. It also is reflected in the general feeling that Congress or the Department of Energy (DOE) should have the ability—in fact, the responsibility—to choose wisely among technologies and determine the rates of resource substitution. Congressional members ask in hearings: "Which synthetic fuel technology should we go with?" The DOE wants to know: "Which biomass conversion technology should we fund for developmental research?" This belief in the need to make such choices on a centralized basis probably makes sense with a technology like nuclear power, which requires massive investment and a national commitment to assure development and use of any one approach, but such a need may be more unique than common.

Agricultural production is probably the most dispersed and diverse of the major economic activities in the United States. Since the diversity of resources and environments is almost infinite, there is a very strong case for not making choices about adjustments to production systems on a centralized basis. The decisions to change, adopt, expand, or go out of business will be made at the farm enterprise level. The more that technology and systems alternatives can originate and be developed in an environment close to the specific farm enterprise, the more likely that these alternatives will be viable ones.

Another major issue, the energy price trade-off between efficiency and equity, is illustrated using irrigation in the Texas High Plains as an example. Should this area go out of irrigation? Should the farmers receive an agricultural exemption that protects them against price increases in the natural gas used as a pumping fuel? There have been several attempts to provide such exemptions. Decisions here will determine the extent to which agriculture will have to adjust to changing energy availability and prices. This, in turn, will influence technology, cropping systems, and the mix of commodities produced 20 to 30 years from now.

AGRICULTURE AS AN ENERGY PRODUCER

The idea of agriculture as an energy producer has aroused a great deal of interest. High hopes have been raised by numerous consultants to the DOE and by others that major contributions will be made to the nation's energy supplies through the production of biomass. The following discussion will be restricted to energy that might be obtained from agricultural production and does not cover energy from forest and related sources. The decision will first consider resource availability; then several specific sources will be discussed.

Food and fiber crops appear to have the edge over energy crops in the competition for land, and this will continue unless the price of energy increases a great deal in relation to the price of food and fiber. Even with tremendous increases in the price of liquid and/or gas fuels, our own domestic coal supplies should provide a long-term cap on the increase in energy prices, which should keep energy crops from displacing food crops.

We do not have an unused land base of any consequence that is suitable for intensive cultivation of energy or food crops (Doering, 1979). Unused land is often unused for very good reasons. Even with increasing demand for food and the expectation of higher food prices, only limited amounts of land could be converted to intensive row-crop cultivation. With government or other subsidies, we could probably convert 30 to 50 million acres from nonfood and fiber uses to biomass production. At the end of 20 years, we would expect that 1/2 of this new row-crop acreage would have been converted from biomass to food production.

It does not make sense to irrigate land to grow energy crops. A number of the large biomass schemes that have been suggested would require irrigation. As a general rule of thumb, water is likely to be increasingly scarce for agriculture and for other uses, so it makes little sense to plan future biomass production based on the necessity of irrigating. Consequently, there is a very definite geographical limitation on the land available for the production of biomass. Most estimates of available land are unrealistic, since they include large areas that are arid or semi-arid.

In terms of looking at specific biomass products and fuels, those feedstocks and processes capable of producing liquid fuels will probably receive priority, since we face more of a liquid-fuel problem than a general energy problem. Alcohol is currently the most popular item in this category and may have been oversold with the claim that alcohol production from agriculture would allow us to thumb our noses at the Arabs and simultaneously solve farm-income problems. It is hard to substantiate fully either of these claims.

Alcohol from Grain

The production of ethanol from corn is already under way in the agricultural sector. It is technically feasible. It is also economic, until the volume of production gets large enough to increase corn prices and/or decrease by-product credits to the point at which costs push ethanol prices over the wholesale price of gasoline. Our current grain ethanol production is around 100 million gallons a year. However, the United States now consumes 100-110 billion gallons of gasoline a year. To achieve a 10 percent alcohol blend on a national basis for gasohol, we would have to produce 10 to 11 billion gallons of alcohol annually. To do this with corn would require 60 percent of the nation's corn crop—with obvious repercussions on the livestock industry, feedgrain exports, and consumer prices. Yet, up to 500 million bushels of corn can be used annually for alcohol production without much adverse impact on markets over the long run. It is probably on this basis that an ethanol industry will get under way.

Alcohol from Cellulose Materials

Within three to five years, we should expect to have commercially proved processes that can convert cellulosic materials, with ethanol as one of the products, in a way that is economically competitive with alcohol from grain. There are a number of potential sources of cellulosic materials, among them crop residues and forage-type materials grown especially for that purpose.

DOE estimates of crop residues available for biomass conversion have been in the range of 400 million dry tons annually. These estimates are the result of driving limited data to illogical conclusions. There has been at least one attempt recently to combine estimates of collectability and appropriate removal based on current harvesting machinery and considerations of soil loss, maintenance of soil productivity, etc. In this case, the estimate is 70 to 80 million dry tons of residue available annually (Bottum and Tynen, 1979).

A major concern in the use of cellulose materials for biomass conversion is that they are bulky and cannot be transported very far before conventional energy sources become more economic. The practical usability of crop residues depends heavily on the density produced. Thus, it is the areas of high-yielding corn, wheat, rice, etc., that will produce the bulk of the usable crop residues for biomass conversion.

Forage is another potential source of cellulose for biomass conversion. It would be possible to increase forage yields on existing forage lands and provide large amounts of cellulosic materials. The great advantage of forage as an energy crop is that it is less demanding of the land base. Table 14 gives an estimate of reason-

Table 14.   Energy from Agricultural Production

| Source | Potential Production | Energy Production[b] | |
		Direct Combustion (quads)	Ethanol (bil. gal.)
Crop residues	39.7-58.7 Mil. tons	.6-.9	1.5-2.3
Crops			
1. From available cropland now cropped[a]	1.95-3.25 Bil. bu.		5.2-8.7
2. From set-aside acres 1978)	900 Mil. bu.		2.4
Forage	68.2-153.3 Mil. tons	1.0-2.3	2.7-6.0
Totals		1.6-3.2[c]	11.8-19.4[d]

SOURCE: Tyner, 1979

a. The figures shown for production from available cropland assume corn is grown on this acreage. The alcohol yield would be about the same, or possibly higher, if sweet sorghum were grown on this acreage.
b. These production figures reflect only technical potential and do not imply that energy production from any of the sources would be economic.
c. Assumes crops would not be used for direct combustion.
d. These totals are somewhat high because some of the land now used for pasture is counted in both the forage and crop categories. With the available statistics, it is impossible to separate clearly these categories.

able potential for biomass production that could be obtained from grains, residues, and forages for conversion to liquids.

It is reasonable to expect that the agricultural production sector could become self-sufficient and even a slight net-energy exporter. However, the sector would not be totally independent unless it could produce the right form of energy and get it to the right place when needed. This would not occur unless the production of energy from agriculture were specifically geared to a closed-loop concept, which would change the overall level and type of production. Energy production from the agricultural sector will certainly not be sufficient to allow us to achieve anything like energy independence on a national basis. The agricultural sector will be only one of a number of contributors necessary to achieve such a goal.

EXPECTATIONS AND PERSPECTIVE

Many expect the magic of technology to offer a remedy that will allow current patterns of energy use to continue with little or no real change required. If that were to happen, the agricultural sector would not become a producer of energy or have to adjust its basic resource-use patterns. Adjustments cause pain, and such pain will only be suffered if it is forced by economic or other circumstances.

The implicit assumption in this paper is that there will be no discovery that takes care of our energy problems for us. Those in agriculture will have to cope by adjusting cropping systems and by adopting what limited new technology actually is practicable. Liquid fuels from coal or shale will not be available in any quantity for at least 20 years, and the short- and intermediate-term adjustments will have to be made largely on the basis of demand management rather than supply enhancement. It is only on such a basis that the resource-use adjustments will be forced to take place and that agriculture might be forced to become a serious producer of energy. Such policies as the continued holding down of natural gas prices reduce the economic incentive for change, at least until rationing becomes necessary and provides another sort of incentive.

If any resource adjustments in agricultural production are to be expected, it is critically important that there be some thinking ahead about the size and type of land base that will be available. Switching from land-intensive to land-extensive or low-energy agricultural systems is not even a possible alternative unless the extensive land base is available. Closed-loop or energy-self-sufficient farms are more feasible on mixed land bases, and we have tended to move toward specialized land bases. We are not likely to shift back to a more traditional mixed land base unless some economic stimulus or government policy reverses the current trends.

The climatic characteristics of our future land base are also going to be critical in determining our future options with respect to the energy intensity of agricultural production and energy production from agriculture. Will the free market for land or government policy act so as to maintain land in agriculture in areas of adequate rainfall? This has not happened in the past. We have developed fringe land and irrigated it at high energy cost to replace or add to adequately watered heartland. It does not appear that the free market will make the energy-conserving choice at current relative prices for energy and different quality land.

Closely related to the land-base question is the trade-off between energy production from agriculture and the environmental load on the land. How much and what kind of production will be expected from a given land base? The general argument that any energy production from agriculture will result in environmental degradation is not well founded. The operative question is whether the expectation for a specific resource base is out of line with its capacity to produce without undue degradation of the land and its associated watershed. This question ultimately will have to be asked at the individual farm enterprise level and decided on the basis of the specific resource endowment and combination of best management practices.

Finally, there are two broad areas of research and development that appear to hold much of the potential for adjustments in energy use and for the production of energy from agriculture. These are photosynthesis and genetics, and information systems. The first is key to the innate productivity of the whole agricultural sector, and the second is key to our ability to fine tune the operation of the sector. Both are critical in helping determine how much production is obtained for the combination of energy and other resources we apply to the agricultural sector. The decision that will need to be made is whether we are willing to invest public dollars to speed up research and development in these areas.

Our approach should be optimistic in spite of the stress given here on the constraints on agricultural production systems. We have made substantial resource adjustments in the past that have been driven by government policy and the economic advantage of substituting less-expensive resources to enhance productivity. Shifts in policies and/or relative resource prices can certainly be the driving force for future adjustments. As in the past, the adjustments may be more gradual than cataclysmic, but gradual adjustments may still take us down a path that is considered to be a radical departure in the light of past experience. Changes do not have to take place suddenly to be important—look at those changes that have brought us to where we are today.

REFERENCES

Doering, O.C. *An Energy Based Analysis of Alternative Production Methods and Cropping Systems in the Corn Belt.* NSF/RA-770125. Washington, D.C.: National Science Foundation, 1977.

Doering, O.C. *Crop Availability for Biomass Production.* Contract Report for the Office of Technology Assessment. Washington, D.C.: OTA, 1979.

Doering, O.C., and Peart, R.M. *Evaluating Alternative Energy Technologies in Agriculture.* NSF/RA-770124. Washington, D.C.: National Science Foundation, 1977.

Eidman, V.; Dobbins, C.; and Schwartz, H. "The Impact of Changing Energy Prices on Net Returns, Production Methods, and Kilocalories of Output for Representative Irrigated Farms." 1975. Mimeographed.

Federal Energy Administration Office of Conservation and U.S. Department of Agriculture, Economic Research Service. *Energy and U.S. Agriculture: 1974 Data Base.* Vol. 1. FEA/D-76/459. Washington, D.C.: FEA, 1976[a].

Federal Energy Administration, Office of Industrial Programs. *Energy Use in the Food System.* Washington, D.C.: FEA, 1976 [b].

Tyner, W.E., and Bottum, J.C. *Agricultural Energy Production: Economic and Policy Issues.* Station Bulletin No. 240 (September, 1979). Lafayette, IN.: Purdue University, Agricultural Experiment Station, 1979.

U.S. Department of Agriculture. *Agricultural Statistics, 1955-1979.* Chapter 9: Farm Resources, Income, and Expenses. Washington, D.C.: USDA, 1955-1979.

U.S. Department of Agriculture. *Agricultural Statistics 1978.* Washington, D.C.: USDA, 1978.

U.S. Department of Agriculture. *Program Report and Environmental Impact Statement.* Review Draft. Washington, D.C.: USDA, 1980.

U.S. Department of Agriculture, Economics, Statistics, and Cooperatives Service. *Farmer's Use of Pesticides in 1976.* Agricultural Economics Report No. 418, pp. 8, 9, 14, 15. Washington, D.C.: USDA ESCS, 1978.

U.S. Department of Agriculture, Economics, Statistics, and Cooperatives Service. *Changes in Farm Production and Efficiency, 1978.* Statistical Bulletin No. 628. Washington, D.C.: USDA ESCS, 1980.

U.S. Department of Agriculture, Soil Conservation Service. *1977 National Resource Inventories.* Washington, D.C.: USDA/SCS, 1978.

Young, K.B., and Coomer, J.M. *Effects of Natural Gas Price Increases on Texas High Plains Irrigation, 1976-2025.* Agricultural Economic Report No. 448. Washington, D.C.: USDA ESCS 1980.

CHAPTER 6

Crop Monoculture and the Future of American Agriculture

JACK R. HARLAN

Few events in recent agricultural history have so greatly impressed the public at large as the epidemic of leaf blight that struck southern corn in 1970. The estimated loss was about 15 percent of the corn crop, or about 700 million bushels. A 15 percent loss of some other crops would be accepted as normal or perhaps even considered good. But we had become accustomed to healthy corn; therefore, an epidemic of such magnitude appeared threatening.

As a result of the leaf-blight epidemic, the public became aware of "genetic uniformity" in crop plants. We learned that such uniformity could render a crop vulnerable not only to disease epidemics but, possibly, to devastation from insects, climatic stress, and other hazards as well. Hence monocultures and genetic vulnerability have become major topics of concern. In this paper, I will present the historical background that has brought American agriculture to its present status. I will also discuss a sample of the most important crops in terms of genetic homogeneity and examine some of the current and projected strategies for dealing with crop monoculture.

HISTORICAL BACKGROUND

The movement toward crop monocultures started with the first tentative efforts at growing plants and rearing animals for human food. Agricultures have evolved many times in many parts of the world, based on plants, and sometimes animals, extracted from the local flora and fauna. Whenever we have adequate archaeo-

logical evidence to trace the evolution of agriculture, we find that
it emerged out of broad-spectrum, hunting-gathering food procurement systems. Such systems can be detected in the Upper
Paleolithic period (around 12,000-10,000 B.C.) or near the close of
the Pleistocene, with some systems surviving into the ethnographic present. By combining archaelogical data with ethnobotani-
cal field work and ethnographic descriptions of the recent past, a
rather consistent picture emerges of these hunting-gathering
economies (Harlan, 1975, and Reed, 1977).

The food sources varied widely in hunting-gathering economies. Different species of plants, mammals, birds, fish, and other
animals were systematically exploited at appropriate seasons.
Groups of people worked out regular schedules for the harvesting
of different resources. Sometimes the resources were sufficiently
abundant to enable people to settle in permanent villages and
extract adequate food from the nearby landscape, as in mesolithic
Europe, the Near East, Jomon Japan (10,000-300 B.C.), and on the
Pacific coast of North America. In other cases, people had to forage over a wider area and move camp at periodic intervals, as in
the Archaic of the Americas, the recent Indians of the Great
Basin the Australian Aborigines, and South African Bushmen. A
large number of species were exploited, and the systems were
basically stable.

With the beginning of agriculture, a small number of plant
species was selected for cultivation. The evolution of agriculture
as a food-procurement system was slow, usually requiring several
thousand years to move from the earliest tentative stages to a
state of dependence on domesticated crops. The transition often
can be demonstrated archaeologically with a change in stone tools
(e.g., production of sickle blades of flint, grinding stones) or by a
change in plant remains recovered from village and camp sites.
Not uncommonly, physical anthropologists can detect the degrees
of agricultural dependence by the teeth and bones in burials. An
increased dependence on cultivated plants often increases tooth
decay and loss (Angel, 1975). Stone-ground grain takes its toll on
teeth; in some populations, teeth are ground off to the gum line
by the age of 35 or so. Lines of arrested tooth growth may signal
serious disease or famine. Studies of osteopathology indicate the
incidence of chronic diseases to be higher among agriculturalists
than among hunter-gatherers. We can detect little evidence of

famine among nonfarmers, but famine has been common among agricultural societies (Lee, 1968).

In subsistence agriculture farming people tend to continue to harvest wild plants and animals for food, but the diversity of the diet is reduced because of the dependence on cultivated crops. The number of species exploited is reduced more or less in proportion to the degree of dependence on crops. This move toward the exploitation of fewer and fewer species began with the earliest stages of plant domestication.

Natural ecosystems are enormously complex and evolve highly dynamic equilibrium conditions. Some biomes are much simpler than others, of course; for example, the tundra or taiga are simple compared to a tropical rainforest. But all natural ecosystems are complex, and their component species have co-evolved over great reaches of geologic time. Their interactions are elaborate and dynamic. The dead are decomposed and replaced by other individuals of the same or different species, individuals that must survive the selection of competition. A kind of stability exists in natural ecosystems because of their complexity and the long co-evolution of component populations, but the stability is due to dynamic equilibria and is in no way static.

Agroecosystems are always much simpler than natural ecosystems. A tropical garden or mixed orchard in subsistence agriculture mimics, to some extent, the tropical flora from which the crop plants were extracted (Anderson, 1954). I have taken plant censuses in gardens in tropical America, the Caribbean, West Africa, and in mixed orchards in the South Pacific Islands. It is not unusual to find more than 40 species of useful plants growing together on a small plot of land. The number of plants of each species is small and there is little apparent system or order. As soon as a plant has served its purpose, it is replaced by something else. These agroecosystems are much less complex than a rainforest, but they too have their balanced equilibra and are generally stable.

The extensive wheat fields of western Kansas superficially resemble the grassland steppe flora they have replaced. This appearance is deceiving, however, for the well-cultivated wheat field is one of the simplest artificial ecosystems in terms of species components. The natural grassland biome of the Great Plains is, by contrast, highly complex, and its steady-state equilibria highly dynamic. This contrast will be discussed to demonstrate that the simpler systems pose greater hazards than the most complex ones.

The move toward monocultures was accelerated by the rise of civilizations in the Near East, Mesoamerica, and South America. One of the characteristics of civilization is urbanization. City people are food consumers, not food producers. The farmers who sell produce to the city tend to grow the surest and most profitable crops for export from the farm. Other crops, therefore, may be discontinued.

The first historical record of monoculture agriculture on a considerable scale comes from Mesopotamia in the second half of the third millennium B.C. Jacobsen and Adams (1958) and Adams (1958) document a shift toward barley monoculture associated with a dramatic decline in crop yield. About 2400 B.C., yields were as follows, based on a reasonable number of records: barley 2,537 liters per hectare (l/ha), emmer wheat 3,672 l/ha, and wheat 1,900 l/ha. The emmer was, no doubt, reported in the glume and would weigh about 75 percent as much as wheat. By 2100 B.C. the yield of barley had dropped to 1,460 l/ha and wheat had almost disappeared. This was due to salinization of the irrigated land resulting from inadequate drainage. This led to the collapse of a number of city-states or small kingdoms in southern Mesopotamia.

The first historical famine on a massive scale caused the collapse of Old Kingdom Egypt (ca. 2160 B.C.). For 170 years no major public building was constructed, and few monuments were erected and inscribed. The few that have survived are starkly eloquent (Bell, 1971, and Breasted, 1906).

> I kept alive Hefat and Homer . . . at a time when . . . everyone was dying of hunger on this sandbank of hell . . . all of Upper Egypt was dying of hunger to such a degree that everyone had come to eating his children, but I managed that no one died of hunger in this [province] [Ankhtifi, chief magistrate of Hierakonoplis and Edfu].

> Plague stalketh through the land and blood is everywhere . . . Many men are buried in the river . . . the towns are destroyed and Upper Egypt is become an empty waste . . . the crocodiles are glutted with what they have carried off. Men go to them of their own accord. Men are few. He that layeth his brother in the ground is everywhere to be seen . . . grain hath perished everywhere . . . the store house is bare, and he that hath kept it lieth streched out on the ground [Bell, 1971].

There is no reason to attribute this disaster to monoculture, for it was caused by a series of years with low floods on the Nile that inundated insufficient land. Nevertheless, there could be a lesson here for us. The Old Kingdom had thrived for a thousand years, safe and secure because of the Nile. Egypt was the granary of the world. Although the desert tribes, the nomads, and the Israelites might be burned out by drought and forced to seek refuge in Egypt, Egypt was assumed to be safe. Yet, the Old Kingdom collapsed, and Egypt starved into obscurity. Today, we are the granary of the world; famine in the United States is unthinkable. We are as secure as ancient Egypt. And that is the point. If it happened to them, could it happen to us? Famine has never been a stranger to agricultural societies.

In ancient civilizations, equilibria often were attained in which about 80 percent of the population supported the whole by agricultural pursuits. Of the whole population, 20 percent could be spared to become shopkeepers, artisans, priests, soldiers, or idle rich. This ratio seems to have been fairly stable and widespread until the industrial revolution, which shaped a completely new world. The degree of industrialization is closely and inversely correlated with the percent of the population engaged in agriculture. Figures for some recent years are shown in Table 1. Living standards are also closely and inversely correlated with the percent of people engaged in agriculture, since wealth generated by industry differs from that generated by agriculture.

With the concentration of more and more people in industrial centers and the relative decline in rural population, it became necessary to devote vast areas to a few staple crops so that urban populations could be supplied with basic food needs. It is these basic staples that are of most concern to us here. Not only the United States and other industrialized nations, but the whole world has drifted in the direction of complete dependence for survival on a small number of food crops. These must, of course, be produced on an enormous scale, thus making monocultures inevitable.

The extent of the drift toward dependence on a small number of food crops on a world scale is shown in Tables 2 and 3. Basic data on gross production were obtained from the 1977 Food and Agriculture Organization (FAO) Production Yearbook. Dry matter data were taken from Morrison's *Feed and Feeding* (1968), and wastage was estimated for each crop from a variety of sources. Rice is reported in paddy, which means it must be hulled for a 20

Table 1. Percent of Population Engaged in Agriculture

	1965	1977
Developed Market Economics	16.5	9.6
North America	5.6	2.8
U.S.A.	5.1	2.5
Western Europe	19.3	11.8
U.K.	3.4	2.2
Oceania	10.3	7.0
Other	26.7	15.3
Developing Market Economies	68.1	60.7
Africa	78.1	71.2
Latin America	44.2	35.9
Near East	65.5	56.2
Far East	70.9	64.2
Other	78.7	73.4
Centrally Planned Economies	60.2	50.5
Asia	71.5	62.6
East Europe and USSR	35.6	22.3
All Developed Economies	23.4	14.1
All Developing Economies	69.4	61.4

SOURCE: FAO, 1977

Table 2. 1977 World Production (millions of metric tons)

	Gross	Est. Edible Dry Matter
All Cereals	1,459	1,215
All Root Crops	570	146
All Pulses	48	43
Vegetables & Melons	319	32
Fruits	257	26
Nuts	3.6	3
Oilseeds (oil equivalents)	45	45
Sugar (centrifuged, raw)	92	87
Totals for All Food Crops Reported	2,793.6	1,597

SOURCE (TABLES 2 and 3): FAO, 1977; Morrison, 1968

Table 3. Estimated Edible Dry Matter of Major Crops of the World (millions of metric tons)

Wheat	348		Beet sugar	32
Maize	315		Cassava	31
Rice	264	1050	Rye	22
Barley	123		Cottonseed	15
Soybeans	70		Peanuts	13
Potatoes	62		Beans	12
Cane sugar	55		Peas	11
Sorghum	49		Coconuts	9
Oats	48		Bananas	8
Sweet potatoes	44		Sunflowers	7
Millets (several)	39		Chickpeas	6

percent loss of weight. Cassava roots are considered about 85 percent usable, but water content also must be considered. The peels were arbitrarily left on the potatoes and sweet potatoes but removed from cassava. Pulses and cereals arbitrarily were considered to have 90 percent dry weight. Although the data vary greatly in reliability, no errors in calculation or reporting are likely to conceal the overall pattern. Humans, as much as any sparrow or canary, have become eaters of grass seeds.

From Table 3, it is evident that the four leading cereals—wheat, maize, rice, and barley—produce more in terms of edible dry matter than all other crops reported by FAO put together. Of course, a large amount of maize and barley is fed to livestock, and production figures are not consumption figures. Even so, the pattern is very clear, and the drift is likely to continue to intensify. The extreme dependence on cereals is, I believe, a relatively late phenomenon historically. The filling of the rice bowls of Asia is so recent that it can be documented by statistics (Table 4). Parts of the lower Mekong Delta are still not settled by rice farmers, but it seems safe to predict that they will be—despite wars, famine,and political instability. The population pressures and ecological imperatives have set the course, and the area in monoculture will increase, not decrease.

And so it is in the temperate world as well. Any new lands left to be opened up are likely to be somewhat marginal, extensive, and the choice of adaptable crops limited. In the United States, the Wheat Belt may be a gigantic distortion of nature, but it is there

Table 4. Expansion of the Area in Rice in Some Countries of Southeast Asia (in 1,000 hectares).

Region	19th Century		Recent		Increase Factor
	Year	1,000 ha	Year	1,000 ha	
Thailand	1850	931	1970	6,727	7.2
Cochin-China	1880	486	1970	2,510	5.1
Burma	1866	708	1970	4,809	6.8
Lower Burma Only	1866	567	1948	2,792	5.0
Java & Madura	1815	848	1960	3,514	4.2
	1900	2,787	—	—	1.3

SOURCE: Grigg, 1974

for ecological and economic reasons, and we are not going to move it. The Corn Belt also will stay where it is, and we will continue to grow corn and soybeans there. The Cotton Belt has moved in recent decades and, if water for irrigation becomes uneconomical on the High Plains, it may move again; yet cotton will continue to be grown in monocultures.

Monoculture is a feature of modern agriculture, and we shall have to learn to live with it; indeed, we might die without it. There are too many people in the world for us to go back to the more complex and more stable agroecosystems of ancient times. Unless we wish to reverse the trend and become less industrialized, unless we are willing to accept lower standards of living, unless we are willing to have fewer people on earth, monocultures of our major crops are here to stay. The question is not whether we shall grow crops in pure stands over large areas; the question is how we can raise the crops on a large scale at minimum risk.

The risks of monoculture have been described many times. Generally mentioned are vulnerability to disease, pest, and adverse environments. Vulnerability is usually equated with genetic uniformity. The relationship is most clearly seen in crops such as wheat, oats, barley, and flax, in which the host and pathogen show intimate and specific genetic interactions. Particularly among the rusts (*Puccinia* species), gene-for-gene relationships occur in which a gene for resistance of the host to a given race of rust is matched by a gene for virulence in the pathogen. Historically, defenses against rust epidemics have taken the following forms: (1) a variety is developed with good resistance to the locally prevailing races of rust; (2) the variety is grown on a large scale, partly because of its resistance; (3) a race of rust capable of attaching to the host evolves, increases rapidly, and causes an epidemic; (4) a gene for resistance to the new race is isolated and bred into a new variety; (5) the variety is grown on a large scale and the cycle starts over again. It is a "Catch 22" with no end in sight.

The selection of a variety of host with constant genotype is, of course, a very efficient screen for isolating a virulent race and providing it with ideal opportunities for buildup. (This strategy does not need to be followed, however; a mix of resistance genes can be deployed, and a high degree of genetic uniformity is not essential to modern agriculture.) The 1972 report by a committee of the National Academy of Sciences provides some details concerning genetic uniformity of major crops as of 1969-70. It is found, for example, that six popular corn inbreds were used as parents of hybrids sown on 71 percent of the acreage. In soybeans, six sources accounted for 56 percent of the acreage sown, and hybrid sorghum production was based on a cytoplasmic sterile system very much like the one in maize that proved susceptible to leaf blight in 1970. Indeed, all of our major crops are considered to have a narrow genetic base and consequently are vulnerable to virulent diseases, insects, or erratic climate. The situation has not changed much since the committe report, but some points that I consider to be important for this discussion were not brought out in the study.

For example, J.A. Clark wrote in 1936:

> More striking than the growth of an oak from an acorn is the fact that the vast red spring wheat industry in the United States, with all the milling, baking, transportation, and trading depedent on it, developed from a few seeds saved from a single wheat plant.

Clark was describing the origin of Marquis, which he considered to be "the greatest achievement in wheat-breeding history." Marquis was, indeed, a remarkable wheat—and was, quite predictably, eliminated because of susceptibility to a race of rust. Its history is instructive. The source can be traced from Galicia in Poland to Germany to Scotland to Canada, where a single plant selection by David Fife of Ontario produced Red Fife. Another Canadian, C. E. Sanders, crossed Red Fife with Hard Red Calcutta from India and selected out Marquis, which later became a parent of Ceres, Hope, Reward, Marquillo, Reliance, Thatcher, Sturgeon, Comet, Tenmarq, and many others. Red Fife was introduced into the United States about 1860 and Marquis in 1912, the one tracing to a single seed and the other to a single cross (Clark, 1936). The elite germplasm was worked and reworked over and over by wheat breeders. Pure lines were selected, crossed with other lines, more lines selected out, and the process repeated. This pattern of usage of germplasm is a model repeated many times over in American agriculture.

But Marquis was no more remarkable than Turkey, introduced into Kansas by Mennonite immigrants from Russia about 1873. A variety survey made in 1919 showed that Turkey occupied 99 percent of the hard red winter wheat area and constituted 30 percent of the total United States wheat acreage. It then was seeded on more than 21 million acres, almost twice that of any other cultivar at the time. (It is worth noting that the most widely grown variety in 1969, as reported by the National Academy of Sciences [1972], was sown on less than 8 million acres and represented less than 15 percent of the total wheat acreage.) The germplasm is still the foundation of the hard red winter wheat industry. More than that, Turkey was one of the parents of Norin 10, the source of the semi-dwarfing genes used in modern high-yielding wheats around the world. The Japanese had obtained Turkey from the United States in 1892 and used it to develop some of their wheats, including Norin 10 (Quisenberry).

The soft wheat belt of the eastern states also was dominated by a single heterogenous source, although not quite as completely. The source was called Mediterranean when it was introduced in 1819, probably from Italy.

The National Academy report on genetic vulnerability was quite correct in pointing out that the genetic base of the wheat crop in the United States is narrow. The fact is that it always has been narrow. Indeed, there is almost certainly more genetic diversity deployed in the wheat crop today than there was 50 years ago. Although it will not be possible here to go into too much detail, I shall attempt to appraise briefly the situation in a select sample of American crops.

GENETIC DIVERSITY OF AMERICAN CROPS

The sample of crops to be discussed (Table 5) is composed of the 10 most valuable crops according to data provided in the 1978 USDA Agricultural Statistics. (These statistics cover the year 1977.) These crops are corn, soybeans, alfalfa, wheat, cotton, tobacco, sorghum, potato, rice, and oats.

Corn

Corn has traditionally been the backbone of American agriculture. It is by far the most valuable of our crops, and its real farm value must be placed between 14 and 15 billion dollars annually.

The vulnerability of corn in the leaf-blight epidemic of 10 years ago was not due to genetic uniformity in the conventional sense but to cytoplasmic uniformity. A special cytoplasm was used for the production of hybrid seed, and a special race of the pathogen evolved that was highly virulent on any corn with this cytoplasm. The problem was corrected rather easily and promptly, although at considerable cost and more expensive hybrid seed. While genetic resistance can probably be developed in the susceptible cytoplasm, seed companies have been understandably reluctant to take the risk, and most hybrid seed is now produced by mechanical detasseling.

Table 5.   Most Valuable U.S. Crops, 1977[a]

Crop	Value $(in billions)
1. Corn	12.9
2. Soybeans	9.9
3. Alfalfa	4.9
4. Wheat	4.7
5. Cotton	4.0
6. Tobacco	1.9
7. Sorghum	1.4
8. Potatoes	1.3
9. Rice	0.9
10. Oats	0.85

SOURCE: USDA, 1978

a. The figures are not all inclusive. The value for corn, for example, is for grain and does not include silage, fodder, sweet corn, or seed corn, and the latter comes close to another billion dollars in value. The figure for sorghum also ignores forage, silage, syrup, and broom-corn. No value is given in USDA statistics for alfalfa. The figures that are provided are quaintly concealed in a chapter entitled "Statistics for hay, seeds, and minor field crops." Here one can find an estimate for hay production of alfalfa and alfalfa mixtures. Even these data have been reported only in the last few years. I have simply multiplied this figure by a reasonable value for a tone of alfalfa hay. To this is added the value of alfalfa seed, which is given by USDA. It may come as a surprise to some that alfalfa is the third most valuable crop in the country—and, of course, it is not always so. Much depends on prices in a given year. The value shown for alfalfa does not include sowings for grazing, but the value for the wheat crop does not include wheat pasture either, which might be enough to bring wheat above alfalfa. For our purposes, it does not matter, but it is strange that better data are not available for one of the country's most important crops.

It is no doubt true that the genetic base of corn is narrower now than it was a century ago, when every farmer was a plant breeder and all varieties were open-pollinated. On the other hand, the national average yield then was well below two tons per hectare, vs. nearly six tons today. Landrace* material is dependable and consistent but low yielding. Genetic improvement inevitably means selection of elite material and narrowing of the genetic base.

The 1970 epidemic demonstrated that the hybrid seed-corn industry has the capability of changing methods of production quickly and that new hybrids can be deployed in a short time. All companies have arrays of inbreds that can be substituted in various combinations in case of an emergency. The larger companies, at least, have facilities in the tropics or in the southern hemisphere where multiplication and/or hybrid seed production can be carried out during our winter season. In a sense, this is a substitute for diversity deployed in the field—the diversity is in the breeding nursery and can be brought forth on call. But this measure of security depends on a strong, independent, and competitive seed-corn industry. Some seed companies have been purchased by international conglomerates, and serious concern has been expressed about the possible development of monopolies or cartels that might reduce competition or the ability of seed companies to move quickly and independently in the event of a crisis. There is reason for concern if the trend continues, but as of now the market share of small, independent seed-corn companies is probably higher than it has been in some time.

At present, corn in the Corn Belt is a rather healthy crop. Plant breeders routinely challenge their materials with a wide array of diseases and pests. A company cannot afford to put out risky hybrids since farmers have many choices and can and do change brands after a poor performance. Profit depends very much on the reputation of the company. On the other hand, this leads to very conservative approaches. The tried and true elite material is used over and over to keep risks at a minimum, and this has the effect of narrowing the genetic base. It is clear that state and federal research has a complementary role to play in developing new and unrelated gene pools for use in the industry. A univer-

---

*A landrace is a mixed population adapted to local conditions of climate, culture, diseases, and pests.

sity can take more risk with less-adapted materials and is under less pressure to put something in the field.

In sum, corn monocultures are not too worrisome, provided we are able to direct our research efforts sensibly and maintain a substantial number of indepedent, financially solvent, competitive seed-corn companies. The number of scientists involved in the breeding and hybrid seed production phases of the industry is considerable. Many of them have had broad experience with the crop not only in the United States but in many other countries, and the industry as a whole can react with speed should an unforeseen threat occur. It is important that this ability to react with speed be protected.

Soybean

The National Academy reported that six varieties accounted for 56 percent of the soybean acreage. The sources of the American crop are given in detail by Hartwig (1973). The most-used varieties trace to introductions from Manchuria in 1911, which produced Mandarin and its derivatives, and in 1926, which resulted in Richland and its derivatives. Soybeans in the north and north central states are almost entirely derived from these sources. In the southern states, additional sources were used, including an introduction from Korea, one from Japan, and one from Nanking (Harlan, 1977). While the narrow base is a risk, it seems that we continue to find useful genetic traits in it, including resistances to insects, nematodes, diseases, and other hazards.

However, soybeans grown in the United States are far removed from a center of origin where they would be challenged by a full array of pests and diseases. Many of these have not yet been introduced to the United States from Asia, but our own soybean varieties would be susceptible to them. The introduction of a pathogen such as soybean rust (*Phakopsora pachyrhizi*) could be devastating. Due to soybean rust, yield reduction of 60 to 70 percent is reported from Australia (Ogle, 1979). Not only are our current varieties of soybean susceptible, we do not even know of a source of resistance in our world soybean collection. High resistance or immunity is known in some wild species of *Glycine* and genetic transfer from these should be investigated urgently. Potential threats should be studied overseas before more diseases and pests are introduced.

Alfalfa

Barnes, *et al.* (1977), have reviewed the problem of genetic vulnerability of alfalfa in some detail. They consider that there have been nine major sources of alfalfa in germplasm introduced to the United States: Ladak, Turkistan, Baltic, Turkish, Flemish, Chilean, Peruvian, Indian, and African. In 1955, about 33 varieties were being grown in the country from one of these sources or from two or three sources in one variety. Since then, a number of seed companies have taken up alfalfa breeding as well as seed production and sales. By 1977, we had about 160 varieties, some with so broad a base that they incorporated germplasm from all nine major sources. Barnes, *et al.* concluded that more diversity is being deployed now than 50 years ago. On the other hand, Barnes *et al.* are not satisfied with the unsystematic collection of germplasm currently available and recommend that it be upgraded. Since alfalfa is the third most valuable crop in the country, it could well be served by more research. A high priority should be given to monitoring diseases and pests abroad.

The spotted alfalfa aphid was introduced in the early 1950s and caused devastating damage, especially in 1956. The situation was alarming at the time; entire stands were killed on thousands of acres, and the economic loss was extensive. Plant breeders and entomologists soon were able to isolate resistant plants from a number of sources. Aided by the introduction of parasitoid wasps that attack the aphid, satisfactory control was obtained. Later, serious problems with the alfalfa weevil cropped up. This kind of periodic crisis can be expected in the future, and we should know much more about other pests before they are introduced.

Wheat

The development of genetic diversity in wheat was discussed above, but the unique character of the North American Wheat Belt must be emphasized. Annually, we plant a wall-to-wall carpet of a single species of wheat from northern Mexico well into Canada, also referred to as the Puccinia Path. Yearly, rust diseases move over 4,000 kilometers along the Puccinia Path from a region where they cannot survive the summer to a region where they cannot survive the winter and back again. This man-made epidemiological system must be rated as one of the marvels of the biological world.

I have been unable to find something resembling this in natural vegetation. Blue grama (*Bouteloua gracilis*) is a species of native grass that does have a somewhat similar range. It builds massive stands over vast reaches of the High Plains from northern Mexico into Canada and infiltrates into the eastern prairies to some extent. Plants are attacked by a number of diseases, including rusts (Harlan, 1976). Almost every plant has a few pustules, especially on senescent leaves, but very little damage is done, and plants are rarely if ever killed. It would be instructive to examine the genetic defenses of such a species. Unfortunately, the problem has not been investigated in detail by plant pathologists. I once studied the genetic diversity of a small portion of the range (Harlan, 1958), and a colleague made a cytological survey of the material (Snyder, 1953). The diversity within the species was astonishing. We did not even sample material from the northern plains, but the material in the south was so variable that we had to conclude that the genetic diversity displayed was on an entirely different order from that deployed by the wheat crop over the same geographic range (Harlan, 1976a).

Several scientists have studied the natural defense of wild and weedy populations of wheat, barley, and oats in Israel (Anikster, 1979, and Browning, 1974). They were able to analyze defenses deployed against identifiable races of rust. These scientists found that if 1/3 of the plants in a population resisted even the most virulent race, the whole population would be protected. The natural defenses included all known forms of resistance simultaneously— general resistance, specific (gene for gene) resistance, tolerance, dilatory resistance (slow rusting), and probably other systems still unidentified (Browning, 1974). The natural defenses developed through co-evolution of the host and pathogen tend to be extremely complex and stable. Defenses developed in breeding nurseries tend to be simple and unstable. The challenge is to make defenses more complex and more like natural defenses and still produce acceptable and marketable wheat.

Cotton

Since World War II, cotton has been raised under a canopy of insecticides. It seemed simpler and cheaper to spray the insects than to manage or breed for resistance. Many of the insects became resistant to poisons, and for environmental and other reasons, it has become increasingly apparent that other means of control must be exploited. In some regions, the pollinators were killed off with the harmful insects so that the cotton populations are largely self-pollinated, resulting in reduced vigor and productivity. Recent research has taken the approach of integrated pest management. This is much more complicated than fixed-schedule spraying and requires familiarity with the pests and their parasites, but the results are already highly encouraging, and more permanent and satisfactory controls appear to be feasible.

As of 1970, three varieties of cotton were planted on 50 percent of the upland cotton area (National Academy of Sciences, 1972). A 1976 survey reported that about 80 percent of the upland cotton area was planted with six varietal types (Duvick, 1977). Exploitation of exotic germplasm has been constrained by the lack of a suitable tropical station for growing and preserving perennial wild and primitive races. The search for resistance to insects is likely to result in increased use of exotic material in the future, with a consequent broadening of the gene base.

Tobacco

The American tobacco crop is largely derived from Virginia types and Maryland broad-leaf sources introduced at an early date. Burleys evolved mostly in Kentucky from the Maryland source (Harlan, 1977). The genetic base is narrow, as with most American crops, but tobacco geneticists have used wide crosses more than those working with many other crops have. Resistance to mosaic virus was inserted into commercial types from *Nicotiana glutinosa* in 1938, and resistance to wildfire disease was transferred from *N. longiflora* in 1947 (Harlan, 1976b). Diseases of seedlings in the seedbeds have been especially troublesome. Research support for a crop considered hazardous to the health will be difficult to fund with public money; yet tobacco is a valuable exportable crop and is one of the most interesting and useful experimental subjects in biology. It lends itself especially to basic research in cytogenetics and cell biology (tissue culture).

Sorghum

Sorghum is a crop developing differently from the general trend. Sorghum's genetic base started as a very narrow one (Harlan, 1977) and has widened steadily through the decades of this century. The conversion program has been especially valuable and is one of the most innovative approaches in the recent annals of plant breeding (Duvick, 1977). Briefly, tropical sorghums are short day plants and are ill adapted to our latitudes. Since short day habit is dominant, the procedure has been to grow exotic accessions in Puerto Rico, cross them to a temperate standard (Martin), and raise the F1's (first generation) in Puerto Rico and F2's (second generation) in Texas. Those that flower and set seed are sent back to Puerto Rico to be backcrossed to the exotic line, the BC1 (first generation, backcrossed) selfed, and the seed sent to Texas for a repeat. After four or more backcrosses, the exotic material is sufficiently adapted for use in a breeding program. Some converted lines have given good performance on their own, and others have been excellent parents.

Hybrid sorghums are produced using a cytoplasmic male sterile system and, theoretically, could succumb to some mutant race of pathogen virulent on this particular cytoplasm. This is a hazard that cannot be ignored, although no difficulties have been encountered to date. Other cytoplasms and other means of producing hybrids are urgently needed (Duvick, 1977).

Potato

The potato has a narrow genetic base in the United States. One introduction, Rough Purple Chili, provided material from which Goodrich selected Goodrich's Garnet Chili in the nineteenth century. It, in turn, yielded some 170 varieties, including an all-time dominant—Russet Burbank (Harlan, 1977). As of 1970, Russet Burbank constituted over 28 percent of the certified acreage. Kennebec and Katahdin contributed 20 percent and 15 percent, respectively. The three together occupied 63 percent of the certified acreage. Further, potatoes are vegetatively propagated, and every plant in a certified field has the same genotype (National Academy of Sciences, 1972). Modern breeding procedures, however, provide some promise of reducing hazards considerably. Methods have been developed to drop routinely the ploidy level from 4x to 2x and raise it again. More exotic material is now being used than before in this effort.

Rice

Some rice varieties in the United States have had remarkable longevity. Caloro endured for fully half a century in California and, as of 1970, it and a derivative, Calrose, occupied about 85 percent of the California rice acreage (National Academy of Sciences, 1972). In the southern states, over 90 percent of the acreage was planted with five varieties. Carolina Gold and Carolina White were grown in South Carolina for some 200 years (Harlan, 1977). This suggests that rice in the United States is not often subject to attack by virulent pathogens with multiple races. The older varieties are finally giving way and newer ones are coming in, but a narrow genetic base has not caused disaster in the United States crop to date.

Oats

The history of oats in the United States parallels that of many other crops in that oats were derived largely from few sources. Red Rustproof came to us from Mexico about 1848. According to Coffman, *et al.* (1961), some 50 varieties could be traced to this source in the United States and probably over 100 worldwide. Kherson was introduced from Russia in 1896 and resulted in a large number of varieties of the Richland type adapted to the upper Midwest. Victoria came from Uruguay and had good resistance to crown rust (*Puccinia coronata*). The resistance was inserted into a number of varieties, and all of them promptly fell to attacks of Victoria blight (*Helminthosporium victoreae*), a pathogen of native grasses so economically insignificant that it had not even been identified and named. Bond came from Australia, and Landhafer from Uruguay via Germany. A number of varieties can be traced to both (Harlan, 1977). In recent years the oat crop has declined in acreage and usage because of falling demand.

Browning, Frey, and their colleagues have outlined a program for systematic development of resistance genes that should be very effective (Frey *et al.*, 1977). Meanwhile, they have released a series of multilines (a mixture of related lines with different genes for resistance) that so far have performed very well. These multilines have been grown on over a half million hectares each year for several years without economic damage from crown rust, historically the most serious disease of oats in the upper Midwest (Frey, *et al.*. 1977). The genetic base of the crop has been narrow in the past, but use of large-scale hybridization, development of multilines, and incorporation of genes from *Avena sterilis* has probably broadened the base considerably. Particularly significant is the objective of developing genetic defenses modeled on the natural defenses of wild oats in their homelands. The philosophy has great merit, should be widely studied, and should be applied as much as possible to other crops (Browning, 1974).

SOME BASIC CONCERNS

A sample of 10 crops is a small one, but I believe these crops are typical. They are extremely important to American agriculture and collectively illustrate the main concerns I shall now discuss. First, the situation is not as simple as many have assumed. It is often claimed that our crops are becoming more and more vulnerable because of greater and greater genetic homogeneity (Harlan, 1972, and National Academy of Sciences, 1972). This may be true of some crops, but the reverse is true of others. Concern has been expressed over the power of a few seed companies that might dominate the trade and that may be purchased by foreign international conglomerates. The hazards, if any, appear to be greatest where hybrid seed is produced—for example, in corn and sorghum—because seed companies control the entire supply of these seeds. The other crops are less likely to be affected by this problem.

It is important to remember that no blanket statement applies to all cases, and the real world is too complex for simple generalities. For the purpose of this discussion, I will make two presuppositions: (1) our major crops will continue to be grown as monocultures; (2) whether their genetic base is narrowing or widening, most crops presently have narrow genetic bases and a considerable genetic homogeneity. I now turn to the question of the causes of genetic uniformity, the precautions that might be taken, and possible future consequences.

Causes of Genetic Uniformity

The reasons for genetic homogeneity are various, more or less interlocking, and mutually supporting. First, a certain degree of uniformity is inherent in varietal development. Only a specific kind of germplasm will give a high performance in a given region. For example, if you grow 10,000 accessions from a world collection in a test nursery, probably no more than a dozen or so will perform well and these will look very much alike. Possibly none of them will perform as well as material you already have developed. High-yielding, elite populations are developed to a large extent by discarding the genetic components that give an unsatisfactory performance. Hence, high performance entails a certain amount of uniformity.

Second, agriculture has been pervaded in recent decades by a pure-line mentality that equates the good with the pure. Seed certification programs and seed labeling laws seem predicated on the idea that mixtures of any kind are bad. These laws were instituted in part to protect the farmer from inadvertently buying low-quality seed. But farmers are much better managers, much better informed, and more selective in their seed purchases than they used to be. Some rethinking and changing of attitudes might, therefore, be in order.

Farmers like uniformity, however, for several reasons. One, there is aesthetic appeal to neat, clean fields of uniform height and appearance. A ragged field is an embarrassment to the modern farmer. Two, if the ears of corn are all the same size and shape, the combine is more easily adjusted and does a better job of harvesting. Similarly, if all the cotton bolls mature at the same time and the plants are the same size and shape, the picker will do a better job, and there will be less loss and less trash in the lint. Uniformity has economic and convenience benefits as well as aesthetic value.

Packers, processors, and marketers also demand uniformity. If tomatoes are all the same size, they will fit in the can better and will require fewer machinery adjustments. The blending of different lots to make a uniform product is much easier and less expensive if the lots are uniform. Customers demand uniformity in flavor, texture, and culinary behavior. If the products are not repeatable, the recipe does not work.

Thus, the pressures for uniformity are intense, continuous, and from all directions. Some of these pressures are relatively trivial and are imposed because the purchaser is in control. In times of scarcity, these pressures can be forgotten very quickly. Uniformity is a convenience, but it is not as necessary as many people seem to think.

Reduction of Risks

Much has been written in recent years about "integrated pest management programs" (Huffaker, 1976, and Levins, 1980). The objective is not so much to eradicate a disease or pest but to manage the agroecosystem in such a way as to prevent economic loss. This requires a good deal of special information and skill. All known methods of control would be used in an integrated system.

In plant breeding, we would deploy the most complex level of genetic defenses that we could achieve practically. We could not reach the levels of complexity of natural ecosystems, but we could improve what is now in place. General resistance, specific resistance, tolerance, and dilatory resistance all would be used simultaneously. Multilines would be one method of integrating these resistances. Stockpiling resistances in adapted populations is another approach now used. Cropping sequences, sanitation, control of alternate hosts, careful scheduling of crops, use of predators or diseases against insect pests, trapping insects, and growing bait crops or crops to feed predators all would be involved in an integrated strategy. In these systems, insecticides or fungicides are only part of the package of control measures. The poisons are used only when and if necessary, based on careful appraisal of potential economic loss (Metcalf, 1980).

Some programs of this kind already have been put into practice with success, but the programs are new and must be improved. We need to examine our agroecosystems more as biologists and ecologists and less as soil specialists, plant specialists, entomologists, or plant pathologists. Fields are ecological units to be managed and guided in the direction of dynamic but stable equilibria. We have much to learn from natural ecosystems and the checks and balances that stabilize them (Odum, 1976).

The following lists some of the measures we could take to reduce risks of growing crops in monoculture. The list is incomplete and different from a list that would be provided by other scientists. The order is more or less operational and does not necessarily indicate relative importance.

Improve Our Germplasm Collections. Although the quality of germplasm collections varies, none is as complete as it should be. There is no excuse for not having excellent collections of at least our major crops; gaps then can be identified and filled in. These germplasm collections are the raw materials from which we extract desirable traits for our breeding programs. The genetic diversity they contain is essential for our agriculture; yet we do not have a satisfactory system for rejuvenation of collections when seed begins to lose viability. See Timothy and Goodman (1979) for a documentary of world corn collections and the kinds of hazards that threaten all collections. For example, while corn is the most valuable crop in this country, the collection maintained by the USDA is very modest in size, and world collections have been taken very casually.

Evaluate Our Collections. Germplasm collections are not very valuable until they have been screened for useful traits. These collections traditionally have been screened by plant pathologists and sometimes by entomologists looking for sources of resistance to diseases and pests. The screenings are often very incomplete since there is a tendency to stop once resistance has been found. Other attributes that should be identified to make our collections useful for plant breeding include: protein and oil quality; adaptations to drought; responses to cold or heat, to waterlogging, or to saline soils; and level of fertility.

Strengthen Our Plant Breeding Programs. Most plant breeding programs could be improved significantly by better facilities and increased support. Some programs lack adequate field space; many have no facility for off-season nurseries and for an extra generation per year; greenhouse space is often inadequate; and cytological or biochemical support may be lacking. Integrated teams of geneticists, plant pathologists, entomologists, and biochemists should participate in crop improvement programs. We should devote more attention to the production of multilines, synthetics, and adapted populations. Instead of a pure-line mentality, we should insert as much diversity as possible into our varieties without impairing performances or market value (Browning, 1974).

Improve Our Monitoring of Diseases and Pests Abroad. More information is needed on organisms that might attack major crops but that have not yet been introduced. With the volume of traffic among countries today, it is unrealistic to assume that the plant quarantine service could intercept all organisms that might be hazardous. Thus we need to know what species are potential threats and where resistances might be found, or, in the case of insects, what predators or insect diseases might keep the pest under control. An offshore quarantine facility would be of great value for studying pests as well as for introducing foreign germplasm safely.

A Strong Seed Industry Is Essential. The benefits of competition among seed companies were discussed earlier. Should a crisis occur, a rapid and effective response by the seed trade is absolutely essential. In the absence of a crisis, greater diversity will be deployed in the field if there are many independent, financially solvent, and competitive companies in the agricultural system. A company tends to sink or swim according to the performance of its product; the result is the production of good and dependable varieties and hybrids. The smaller companies must depend more than the large ones on inbreds and gene pools developed by state or federal agencies. A strong, independent seed-corn industry would be supported best by stronger corn research programs at the state and federal levels. If there is a danger from company take-overs and mergers, the risks would be reduced most by stronger public research and development.

Greater Tolerance in Marketing and Processing. A high degree of diversity in marketing and processing could be tolerated without an excessive increase in price. New and more versatile machinery could be devised and, no doubt, will be. The problem of designing equipment with greater ability to handle variation in produce size, shape, etc., has probably not been addressed with sufficient focus. It has been easier to adapt the plant to the machine rather than the machine to the plant. The easiest way, however, is not necessarily the best.

More Effective Leadership in the USDA. Coordination on a regional and national scale is essential in the event of an epidemic. The ability of the USDA to respond to a crisis has been greatly reduced as a result of recent organization. The chain of leadership should go through scientists knowledgeable about individual commodities rather than through business offices and voucher specialists.

Expedite Development of Safe Agricultural Chemicals. I do not propose to set aside the Environmental Protection Agency. We have been too careless in the past with poisonous materials. But the time and money now required to develop a useful chemical and demonstrate its safety is so great that new and better chemicals are not forthcoming. Although I do not have a solution to this problem, a serious effort by those directly involved is necessary to reach a solution.

Broaden the Training of Scientists in Universities. Looking at a field as an ecosystem requires a special pair of eyes. If we are to manage these fields and the pests and diseases in them, our agricultural scientists will need more extensive training in biology and ecology and a better understanding of the dynamics of plants' interactions with fungi, bacteria, viruses, insects, nematodes, and the like (Huffaker, 1976, and Levins, 1980). Funding in biology in recent decades has been so heavily tilted toward biochemistry and cell biology that the fields of interaction among species have been neglected. A revised orientation is needed.

All of the above recommendations are easily within our means to effect. Whether or not they will be executed depends too much on decisions made by people who know little or nothing about agriculture.

Future Problems

Finally, the future. How will monocultures constrain American agriculture in the future? It is much safer to look backward than forward. But if history points the way at all, it should be possible to reach some reasonable conclusions.

Our major crops will continue to be grown over large areas in pure stands. We shall have crop monoculture in the foreseeable future. The genetic bases of these crops primarily are narrow, but they always have been narrow. A restricted genetic base is nothing new to American agriculture.

We are aware of the problems and the risks, perhaps, as never before. We are disillusioned by resistances that break down when races of a pathogen shift and by resistances that build up as we spray poisons for insect control. We are disappointed by a treadmill that must crank out a new variety every three to five years in order to maintain a level of resistance that does not cause economic loss. We are dismayed by poisons we have to put in the environment and by the hazards some have caused. On the other hand, we are frustrated by the time and expense required to develop safe and useful agricultural chemicals. We are, I think, ready to take a new look to see if there is not yet "a more excellent way."

Personally, I am optimistic. The components of better disease and pest control systems exist. We need to gather more detailed information, to pursue vigorous programs of basic and applied research, and to react quickly when problems threaten. We may expect our greatest problems and severest losses to occur in those crops that are grown most extensively over large areas such as wheat, barley, corn, sorghum, and soybeans. Monocultures will be one of the causes. We have had disease epidemics and insect outbreaks in the past, and we will have them in the future. We have recovered from them in the past; we shall recover from them in the future. I hope and think, however, that we have learned enough from past mistakes to avoid many epidemics in the future. I see no reason why American agriculture cannot continue to produce more as the demand grows and without "undesirably" higher costs.

If the unthinkable should happen, if our crops should fail and starvation stalk the granary of the world, I am persuaded that the causes would be more likely due to erratic climate than to monoculture as such. Failure of rains in the African highlands caused the eclipse of Old Kingdom Egypt. Too much rain caused starvation on a vast scale in Europe in 1315-16. We have less defense against these calamities than we have in managing genetic vulnerability.

REFERENCES

Adams, R.M. "Salinity and Irrigation Agriculture in Antiquity." Diyala Basin Archeological Project, Progress Report, June 1957 to June 1958. Chicago: University of Chicago, 1958.

Anderson, E. *Plants, Man, and Life.* London: A. Melrose, 1954.

Angel, J. Lawrence. "Paleoecology, Paleodemography and Health." In *Population, Ecology and Social Evolution,* edited by Steven Polgar, pp. 167-90. The Hague: Mouton, 1975.

Anikster, Y., and Wahl, I. "Coevolution of the Rust Fungi on Gamineae and Liliaceae and Their Hosts." *Annual Review of Phytopathology* 17(1979): 367-403.

Barnes, D.K.; Bingham, E.T.; Murphy, R.P.; Hunt, O.J.; Beard, D.F.; Skrala, W.W.; and Tenber, L.R. "Alfalfa Germplasm in the United States: Genetic Vulnerability, Use, Improvement and Maintenance." USDA Technical Bulletin No. 1571. Washington, D.C.: USDA, 1977.

Bell, Barbara. "The Dark Ages in Ancient History. I. The First Dark Age in Egypt." *American Journal of Archaeology* 75 (1974): 1-26.

Breasted, J.H. *Ancient Records of Egypt.* Chicago: University of Chicago Press, 1906.

Browning, J.A. "Relevance of Knowledge About Natural Ecosystems of Development of Pest Management Programs for Agro-ecosystems." *Proceedings, American Society of Phytopathology* 1 (1974): 191-9.

Clark, J.A. "Improvement in Wheat." In *Yearbook of Agriculture, 1936,* pp. 207-302. Washington, D.C.: USDA, 1936.

Coffman, F.A.; Murphy, H.C.; and Chapman, W.H. "Oat Breeding." In *Oats and Oat Improvement,* edited by F.A. Coffman. Madison, WI.: American Society of Agronomy, 1961.

Duvick, D.N. "Major United States Crops in 1976." In *The Genetic Basis of Epidemics in Agriculture,* edited by P.R. Day. Vol. 287, pp. 86-9. Annals of the New York Academy of Science. New York: N.Y. Academy of Science, 1977.

Frey, K.J.; Browning, J.A.; and Simons, M.D. "Management Systems for Host Genes to Control Disease Loss." In *The Genetic Basis of Epidemics in Agriculture,* edited by P.R. Day. Vol. 287, pp. 255-74. Annals of the New York Academy of Science. New York: N.Y. Academy of Science, 1977.

Grigg, D.B. *The Agriculture Systems of the World: An Evolutionary Approach.* London: Cambridge University Press, 1974.

Harlan, J.R. "Blue Grama Types from West Texas and Eastern New Mexico." *Journal of Range Management* 11 (1972 [a]): 84-7.

Harlan, J.R. "Genetics of Disaster." *Journal of Environmental Quality* 1 (1972 [b]): 212-15.

Harlan, J.R. *Crops and Man.* Madison, WI.: American Society of Agronomy, 1975.

Harlan, J.R. "Diseases as a Factor in Plant Evolution." *Annual Review of Phytopathology* (1976 [a]).

Harlan J.R. "Genetic Resources in Wild Relatives of Crops." *Science:* 16 (1976 [b]): 329-33.

Harlan, J.R. "Gene Centers and Gene Utilization in American Agriculture." *Environmental Review,* March 1977, pp. 26-42.

Hartwig, E.E. "Varietal Development." In *Soybeans: Improvement, Production, and Uses,* edited by B.E. Caldwell, R.W. Howell, R.W. Judd, and H.W. Johnson, pp. 187-210. Madison, WI.: American Society of Agronomy, 1973.

Huffaker, C.B. and Messenger, P.S., eds. *Theory and Practise of Biological Control.* New York: Academic Press, 1976.

Jacobsen, T. and Adams, R.M. "Salt and Silt in Ancient Mesopotamian Agriculture." *Science* 128 (1958):1251-8.

Lee, R.B., and Devore, I. *Man the Hunter.* Chicago: Aldine, 1968.

Levins, R. and Wilson, M. "Ecological Theory and Pest Management." *Annual Review of Entomology* 25 (1980):287-308.

Metcalf, R.L. "Changing Roles of Insecticides in Crop Protection." *Annual Review of Entomology* 25 (1980):219-56.

Morrison, F.B. *Feeds and Feeding.* 22nd ed., abr. Clinton, Iowa: Morrison Publishing Co., 1968.

National Academy of Sciences. *Genetic Vulnerability of Major Crops.* Washington, D.C.: National Academy of Sciences, 1972.

Odem, E.P. "Harmony between Man and Nature: An Ecological View." In *Beyond Growth—Essays on Alternative Futures,* edited by W.R. Burch and F.H. Borman, pp. 43-55. Yale University School of Forestry and Environmental Studies Bulletin 88. New Haven: Yale University, 1976.

Ogle, H.J.; Byth, D.E.; and McLean, R. "Effects of Rust (Phakopsora Pachyrhizi) on Soybean Yield and Quality in Southeastern Queensland." *Australian Journal of Agricultural Research* 30 (1979):883-93.

Quisenberry, K.S., and Reitz, L.P. "Turkey Wheat: The Cornerstone of an Empire." *Agricultural History* 48 (1974):98-114.

Reed, C.A., ed. *The Origins of Agriculture.* The Hague: Mouton, 1977.

Snyder, L.A., and Harlan, Jack R. "A Cytological Study of Boutelona Gracilis from Western Texas and Eastern New Mexico." *American Journal of Botany* 40 (1953):702-07.

Timothy, D.H., and Goodman, M.M. "Germplasm Preservation: The Basis of Future Feast or Famine Genetic Resources of Maize—an Example." In *The Plant Seed: Development, Preservation and Germination,* edited by I. Rubenstein, R.L. Phillips, C.E. Green, and B.G. Gengenbach. New York: Academic Press, 1979.

U.S. Department of Agriculture. *Agricultural Statistics, 1978.* Washington, D.C.: USGPO, 1978.

United Nations, Food and Agriculture Organization. *Production Yearbook, 1977.* Rome: FAO, 1977.

CHAPTER 7
Climate Change and the Future of American Agriculture

ROBERT H. SHAW

> The increasing realization that man's activities may be chang-
> ing the climate, and mounting evidence that the earth's cli-
> mate has undergone a long series of complex natural changes
> in the past, have brought new interest and concern to the
> problem of climate variation.
>
> *Understanding Climate Change*
> National Research Council (NRC) Report, 1975

A cooling trend since 1940, catastrophic drought in the African
Sahel, and several poor crop years in different parts of the world
in the early 1970s have centered much attention on possible climate changes and their potential effects on agriculture. Many
questions have been raised. How fast have past climate changes
occurred? What has caused the present cooling trend? Will
increased carbon dioxide cause a warming? Are we really expe-
riencing a climate change? Will cooling continue, or will it
reverse? Answers remain unclear to all these climatological
questions.

What is not in doubt is that climate greatly affects the location
and productivity of agriculture. The United States has large areas
with unusually favorable climate, but even in these areas variations in climate can dramatically affect yields. In marginal climatic
areas, yield levels are particularly sensitive to climatic variations.
Bryson (1980) cites two examples of how climate has helped shape
the history of the Great Plains. The Indians of the northern part
of the Great Plains once lived in scattered villages and grew corn.
Around 1200 A.D. these villages were abandoned due to a long
period of drought in the area. The Pueblo Indians of the Mesa

Verde abandoned their homes about the year 1300 after having successfully maintained an agricultural society for several centuries. Again, it was a period of severe droughts that caused them to migrate.

The question logically can be raised whether these events could happen under modern-day technology. The answer is that, although technology has provided us a buffer against aspects of the weather, we remain very susceptible to extreme weather and climate events. For example, in the 1930s drought and heat wiped out extensive areas, i.e., the "Dust Bowl." In the late 1960s and early 1970s, the average corn yield over the Corn Belt varied over 15 bushels per acre (on individual farms over 150 bushels per acre)—not an occurrence to cause a migration, but one that results in a very unstable level of crop production. In spite of our high level of technology, areas of Iowa had a complete failure of the corn crop in 1977 due to a drought as severe as those of the 1930s. In 1980, the Corn Belt has been ravaged by June hailstorms and excessive rainfall. Crop failures might take place if severe enough weather occurs.

All crop failures, however, are not caused entirely by the climate. The six-year Sahelian drought, for example, was complicated by other factors. The multiplication of drilled wells of high flow yield in the Sahel had permitted concentrations of animals up to 10 times the carrying capacity of the land. According to the United Nations Food and Agriculture Organization (FAO), "Nobody thought that when water is given to cattle, and when so many animals are concentrated in one particular spot it is necessary to supply these animals with food" (Ceres, 1977). Landsberg (Roberts, 1978) adds, "I wish people would quit blaming climate or climate change for failure of agriculture in marginal lands. It is a scapegoat for faulty population and agricultural practices."

The tendency is to accept the good weather as normal, and bad weather as abnormal. In marginal areas, we are overloading a system with limited capacity. During a period of relatively good years, the system may carry this overload, but when a period of bad weather occurs, the system cannot carry this overload, and panic sets in. The food crisis is often people-made—not climate-made. In some climates, periodic droughts are a normal event and must be expected. Although we are certain that droughts will reoccur in the Sahel or other dry areas, we cannot say when they will occur.

This paper briefly discusses what our climate has done in the past and the possible impacts of peoples' activities on the climate of the future, and presents some estimates of possible impacts of climate change on agricultural production. The reader should recognize that the area of climatic change is very complex. The 1975 NRC report *Understanding Climate Change* states this very simply: "Our knowledge of mechanisms of climate change is at least as fragmentary as our data." In this paper, I will discuss "what might happen" rather than "what will happen."

CLIMATE TRENDS, PAST AND PRESENT

How we rank our present climate in terms of history depends on the period of reference used. In the last 500 million years, for example, the earth has been free of polar ice more than 90 percent of the time. Using this as a reference point, we are living in an ice age, an anomaly for our planet (MIT, 1971). In the last million years, however, the earth's climate has undergone an alternation of glacial and interglacial periods, marked by the enlargement and reduction of continental ice sheets in the Northern Hemisphere and by periods of rising and falling temperatures in both hemispheres. At present, we are in an interglacial period; thus, using the climate of the last million years as a reference point, we are in a relatively warm period. Since it is during this interglacial period that civilization and agriculture as we know them have developed, we will use the last million years as our term of reference for this paper.

The main trends in global climate for the last 100, 1,000, 10 thousand, 100 thousand, and 1 million years are shown in Figure 1. In the last thousand years, relatively few periods have been as warm as the present. Our climate is much warmer than, for instance, the Little Ice Age that occurred from about 1400 to 1700 A.D. Historical literature is filled with references to the adverse effects of cold climate on agriculture during that period and the difficulty governments had feeding people.

In his book *Times of Feast, Times of Famine* Ladurie (1974) questions the validity of these broad statements on the detrimental weather of the Little Ice Age and observes that many statements are not documented. He reports that the Little Ice Age had short periods of weather detrimental to agriculture as well as short periods of weather that were relatively favorable.

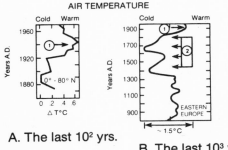

A. The last 10^2 yrs.

B. The last 10^3 yrs.

LEGEND:
1. Thermal maximum of 1940s
2. Little Ice Age
3. Younger Dry as Cold Interval
4. Present Interglacial (Holocene)
5. Penultimate Interglacial (Eemian)

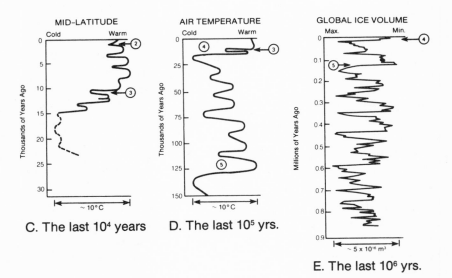

C. The last 10^4 years   D. The last 10^5 yrs.

E. The last 10^6 yrs.

SOURCE: Federal Council for Science and Technology, 1974

(A) Changes in the five-year average surface temperatures over the region 0 - 80 N; (B) Winter severity index for eastern Europe; (C) Generalized northern hemispheric air-temperature trends, based on fluctuations in alpine glaciers, changes in tree-lines, marginal fluctuations in continental glaciers, and shifts in vegetation patterns recorded in pollen spectra; (D) Generalized northern hemisphere air-temperature trends based on mid-latitude sea-surface temperature, pollen records, and on worldwide sea-level records; (E) Fluctuations in global ice volume recorded as changes in isotopic composition of fossil plankton in deep sea core V23-233. Reproduced from Report of the GARP Panel on Climatic Variation (submitted to USC/GARP).

Figure 1. Main Trends in Global Climate: The Past Million Years

Ladurie's records show a wet period just before the Little Ice Age. A very wet decade began in 1310; the floods of 1315 inundated harvests, vineyards, and seeds in the ground, making the 1316 wheat harvest very poor. In France, Flanders, Germany, and England, the poor died by the millions from hunger and epidemics. The food problem was aggravated during this period by the lack of labor due to depopulation by epidemics, famine, and wars. This wet period was followed by a period with longer winters, later wine harvest dates, and longer glaciers. Thus developed the hypothesis that from 1540 to 1600 a very cool period existed.

Overall, cooling during this period averaged only about 1°C below normal. Although the climate was more detrimental than favorable for agriculture during this period, the effects of the detrimental weather were compounded by factors other than climate.

This cool period was the antithesis of the climate from 1850 to 1950. The Northern Hemisphere mean temperature increased 0.6°C from 1880 to 1940, reaching a peak around 1940. Reports from around 1940 expressed concern about the warming trend we were in and the possible disasters that might occur. Since the 1940s, the temperature has fallen about 0.3°C, and concern is now being expressed about the cooling trend we are in. Some studies, for example, Moran, *et al.* (1977), suggest that the cooling trend already may be ending.

The significance of these historical temperature patterns to the future of agricultural development cannot be overstated. The examination of past temperature patterns reveals an important fact: *agriculture as we know it has developed during an abnormally warm period in recent climatic history.*

Scientists do not completely agree on how rapidly past changes have occurred. Bryson (1974) states that ice-age climates (changes of 4° to 6°C) may end (and probably start again) in a century or two, although glacial and oceanic response and a new equilibrium may take a millenium. He also states that, during the past 10 millenia (the Holocene), climatic changes of an order of magnitude smaller (0.4 to 0.6°C) have occurred in decades. These small changes in climatic variables may produce significant environmental changes. Evidence in literature indicates that several interglacial periods began with an abrupt termination of an intensely cold, fully glacial period. Recent inferences drawn by Lamb and Woodroffe (1970) tend to support a time scale of 1,000 to 5,000 years from the transition to extensively glaciated conditions.

Warm periods similar to the present one have occurred in the past; these periods have not lasted longer than 10,000 years. Since the current warm period has lasted almost that long, the climate might be due to change (Mitchell, 1977). If the climate does change, we would have to readjust our lifestyle extensively.

Not only do scientists disagree on how quickly climatic changes occur, they also disagree on the definition of climatic change versus fluctuation.* Suppose, for example, the current warm period was considered, not a climate change, but a more or less random fluctuation of climate. If this is the case, we should expect the present abnormally warm climate to approach more closely the long-term average, i.e., a much cooler climate. That, however, also could pose problems. What if exogenous factors are causing the present cooling trend? We need to understand those factors, to predict how the climate might change in the future. A statement from the 1975 NRC is very pertinent, "There is no doubt that the earth's climates have changed greatly in the past and will likely change in the future. But will we be able to recognize the first phases of a *truly significant climate change* when it does occur?" Unless someone can devise a cause-effect relationship for climate change, we probably will not recognize the first phases of such a change. We will not be able to differentiate between a real change or simply a random fluctuation; we will only be able to guess about climate in subsequent years.

Possible Misinterpretation of Climate Trends

We must be very careful not to misinterpret the facts of climate change by using the wrong historical example or by inferring incorrectly from it. For example, using a short time period as a reference point may well lead to a wrong conclusion. Data of Starr and Oort (1973), for instance show a five year declining temperature trend for the Northern Hemisphere from 1958 to 1963. If such a temperature trend continued, we would quickly find ourselves in another ice age. No such long-range trend was proposed by the authors, but these data have been used by some to indicate a possible climatic disaster.

---

*Landsberg (1975) proposed that a climate change involve a time scale of 10^4 – 10^6 years, a climatic fluctuation a time scale of $10 - 10^3$ years, and that climatic iterations be considered variations occurring in a period of less than 10 years.

People tend to overreact to a climatic problem because they do not view it in the proper historical perspective. The problem may be a short-term event that will revert to a more normal condition, or it may be a rare event that is not indicative of climatic conditions. In addition, short-term climate anomalies tend to be, geographically, very area-specific. In fact, there may be compensating trends occurring elsewhere. Van Loon and Williams (1976) studied Northern Hemisphere monthly temperature data and generally found that when the temperature was increasing in one area, it was decreasing in another area. While the growing season in England has been getting shorter by two weeks (1960s compared to 1940) (Davis, 1972), it has been increasing in length in parts of the United States (NRC, 1976).

Discussions of climate change often involve averages over time and over large areas. These averages can be misleading, especially if applied to specific geographical areas. For example, one report states that the mean temperature changes that have occurred since 1940 are misleading, since in the last 30 years, the temperature changes in Arctic regions were three times those observed in mid-latitudes, with little change in the tropics (MIT, 1971). Similarly, it would be misleading to generalize about the effects of these climate changes on agriculture. The climate in the major areas of grain production in the middle latitudes has changed relatively slightly. The large cooling that has occurred in the Arctic regions is in an area with no grain production.

Climatic changes also may differ for different seasons. Budyko's and Vasischeva's (1971) theoretical calculations show a much greater change in temperature at high latitudes in the cold half of the year than in the warm half of the year with no seasonal temperature changes induced at 50° latitude, assuming the melting of the sea ice in the Arctic. Since much of the world's grain is produced in the 35°-50° latitudes during the summer period, there would be little impact on agriculture. Yarger (private communication, 1976) analyzed temperature data for Iowa and found for the 1940-74 period that only January and the annual temperature show a statistically significant cooling trend. The agriculturally important May-September period showed no significant cooling trend. However, the period analyzed did not include 1934 and 1936, which had very hot summer temperatures. With these years included, resulting in the data series starting at a very high temperature level, Thompson (1979) presents data that do show a summer cooling trend, although this was not tested statistically. Studies that project the possible impact of climate change on agriculture often use the same change for all periods of the year. For high and middle latitudes, this may be an incorrect assumption and could significantly affect yield projections.

My viewpoint on climate change is similar to that of Mitchell (1977), who said, "The pace at which climate change might occur is not something that worries me—but it does some people. It's the more rapid wiggles of climate that concern me".

From the standpoint of agriculture, it is the short-term fluctuation around the climatic norm that is of greater importance. Long-term changes offer some chance for technological change; short-term "climatic iterations" do not. These fluctuations may have little relationship to any cycles or trends in the climate and could possibly be only year-to-year variations of a somewhat unsteady climate system.

Climate Cycles

Up to now, I have primarily discussed trends, not cyclic patterns, in the weather. A number of weather cycles corresponding to solar cycles have been hypothesized. The only one that I will discuss is the so-called 11-year cycle, or, more precisely, the Hale double-sunspot cycle of 22 years in which an 11-year cycle of large amplitude (major) and an 11-year cycle of small amplitude (minor) are combined. From 1610 to 1914, the cycle averaged 22 years; since 1914 the cycle has averaged only 20 years.

A sunspot is an area of very strong magnetic field on the sun. A sunspot cycle starts with only a few spots present. The number of spots increases for 3-5 years, then decreases back to near zero, the

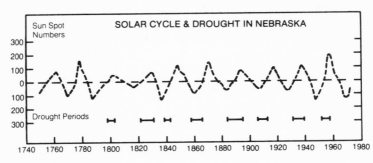

SOURCE: Thompson, 1973

Figure 2.   Drought Periods in Nebraska Plotted against Double Sunspot Cycle

minimum for that cycle. In Figure 2 (Thompson, 1973), the portion of the curve above the zero line is the major cycle and, that below the zero line is the minor cycle. The polarity of the sun's magnetic field is reversed for the two 11-year cycles.

Thompson has shown how the drought periods in Nebraska, as determined by three-ring studies of Weakly (1962), corresponded in phase with the sunspot cycle. Major droughts occurred after the peak of the minor cycle near the time of minimum sunspot numbers. Based on information of this type, Thompson projected severe drought in the Corn Belt in the mid-seventies, an event that did occur. Recently, Mitchell, *et al.* (1979), using three-ring data, have shown that major droughts in the Great Plains follow a 22-year cycle. They found a tendency for the amplitude of the 22-year drought rhythm or cycle in the United States and suggest that the drought rhythm is in some manner controlled by long-term solar variability directly or indirectly related to solar magnetic effects. The solar control is best described as a modulation of terrestrial drought-inducing mechanisms, such that it alternately encourages, then discourages, the development of major continental droughts that are set up by evolutionary climatic processes unrelated to solar activity.

Hancock and Yarger (1979) found statistically significant relationships between the 22-year double-sunspot cycle and January temperatures in the eastern United States. There is also evidence for a significant relationship between June temperatures and the 22-year double-sunspot cycle in the western Corn Belt and midsection of the Great Plains. With the 11-year cycle, significant relationships were found for April precipitation in the southern states from Texas east, May temperatures in a six-state area centered around Ohio, June temperatures in the southern Corn Belt, July temperatures in the area between Texas and Indiana, and June precipitation from Texas to Virginia. Other scattered areas of significant correlation exist. No significant relation was found for July precipitation and August temperature or precipitation.

Weinbeck (1980) performed mathematical analyses on 41 state-averaged monthly temperature and precipitation amounts for 1890-1977. Significant periodicities were observed for wet and dry June precipitation and for June above-normal temperatures. Above-normal January temperatures also showed the same periodicities, particularly along the eastern seaboard. Low precipitation and high June temperatures peaked about two years before the minor minimum, and wet periods tended to occur near the minor maximum but were not as pronounced. No significant cycles were found in July or August temperature or precipitation.

There are thus some indications of cyclic weather patterns that could significantly affect agricultural production. One major problem with projecting from cyclic patterns is that even if a period of years can be shown to have a relatively high probability of an event occurring, the weather for an individual year within that period cannot be projected, i.e., one or more years within a general dry period could be wet, and it is not possible to predict which years. Thus, cyclic patterns provide information that may be of little benefit to the individual farmer, but could be very valuable in national planning relative to farm programs and food reserves. I wholeheartedly agree with Roberts' (1978) statement, "If research confirms the existence of a 22-year drought cycle for the Great Plains (or other areas) this should be seen as part of the 'normal' climate regime, not an anomalous event." Drought is part of the normal climate, even though we cannot accurately predict the years of occurrence or the severity of the drought. We also should recognize that areas can have serious droughts in years other than those indicated by the 22-year cycle, i.e., 1947 in parts of Iowa and Missouri and 1960-62 in parts of Missouri. It is generally believed that off-cycle droughts will not be as extensive or as severe as those indicated by the 22-year cycle.

Variability of Climate

The climate of the earth is now known beyond a doubt to have been in a more or less continued state of flux on all resolvable scales of historical and geological time. (Science Council of Canada, 1976). The current interest in variability concerns short-term variations. Statements such as "Climatologists agree that future climates will have more variability than the climate of the past one or two decades" have appeared in the literature in recent years. Seldom are such statements documented or the type of variability defined. Three types of variability in the climate are important to agriculture: (1) daily, or short-term variability, (2) variability among months or seasons within the same year, and (3) variation between years. The third is probably the type of variation referred to in the preceding statement.

The author has used a biological soil moisture–stress model to show that the droughts of the 1970s in Iowa in the Corn Belt were not nearly as severe as those of the mid-1930s. The analysis shows that droughts in 1934 and 1936 were by far the most severe in the last 40-50 years (intensity and area coverage). The 1977 drought was almost as intense in some local areas as the 1936 drought, although it was not nearly as hot or extensive.

Other than using models, we often have few quantitative measurements available for comparing different periods of detrimental weather. A report of the Bellagio Conference (Rockefeller Foundation, 1975) came to the following conclusions:

- Climatic variability—region by region and from year to year in particular regions—is and will continue to be great, resulting in substantial variability in crop yields in the face of increasing global food needs and short supplies.

- There is some cause to believe—although it is far from certain—that climatic variability in the remaining years of this century will be even greater than during the 1940-70 period.

A report of the Institute of Ecology–The Charles F. Kettering Foundation (1976) stated, "But there is evidence that . . . an unprecedented run of good growing weather in the 1950s and 1960s played an important part in the high yields of U.S. agriculture during that period."

A report of a National Oceanic and Atmospheric Administration Committee (NOAA, 1973) uses Thompson's type regression analysis (1969) and actual weather data over an 80-year period to estimate corn yields for the five major corn producing states. The analysis assumed the 1973 technology time trend value for all years. This trend value estimates what the yields would have been for each of the 80 years, assuming 1973 technology had been used. The analysis was updated through 1975 (NRC, 1975). The results shown in Figure 3 indicated that the 18 years ending in 1973 were all relatively favorable for corn production in the five-state area, i.e., it was a period of relatively small variability in weather in relation to corn yields. No other period appears to be as favorable, although there was another relatively favorable period around 1900. Since 1955-1973 was so unusual in terms of weather history, it is unlikely that it will reoccur, and a more variable period of weather might be expected in the Corn Belt. The NOAA calculations indicate roughly a ±10 percent yield variation due to weather departing from the normal. Thompson has pointed out that this procedure will give estimates in the right

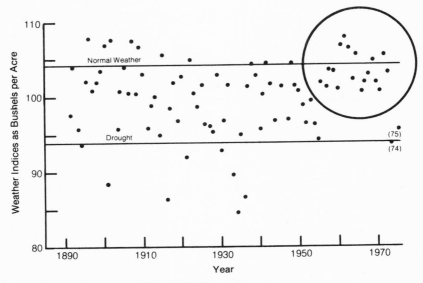

SOURCE: National Research Council, 1976

Figure 3.   Simulated 5-state Weighted Average Corn Yields using 1973 Technology and Harvested Acreage: Ohio, Indiana, Illinois, Iowa, Missouri

direction (plus or minus) but will underestimate the variability due to weather. Comparable values for other crops and areas are not available, but I would expect greater variability on crops grown in more marginal climatic areas. Calculation of the yields since 1975* showed yields of 103, 100, 106, and 105 bushels per acre for 1976 through 1979, assuming 1973 technology. Only because of the yields of 1974 and 1975 do these years show a greater variability than for the 1955-73 period. Maintenance of a stable level of production in the Corn Belt is almost impossible if high weather variability occurs, even with high-level technology.

A similar type study (NOAA, 1973) for the Wheat Belt showed the most recent six years were very favorable for crop production. A number of similarly favorable periods also were evident, indicating that the weather in the Wheat Belt of the United States in recent years has not been that unusual.

Decker (1977) reported on studies in which he had computed the variance of climatic elements by decades computed for different agricultural production areas of the world. The annual precip-

---

*Regression equation provided by Dr. L.M. Thompson, Iowa State University.

itation in the winter wheat producing regions of the Soviet Union did not show a striking decrease in the year-to-year variability during the most recent decades, as was evident in the Corn Belt. The May plus June precipitation in the Canadian wheat region and in the Soviet winter wheat region showed variation less than average in the recent decades, but the tendency for a lower variance did not appear to depart from that expected by normal variability. There was a period of extremely high variability in the Canadian precipitation in the early 1900s. The Soviet area showed a period of high variability between 1925 and 1950.

Decker stated two points that were evident from his analysis: (1) world climate was not unusually low in variability everywhere during the 1955-70 period, and (2) probably more importantly, periods of one or more decades in length with high (and low) year-to-year variability in climate do occur. This type of variation can be extremely important to agriculture, whether due to climate change or simply natural (random) variation. According to the NRC (1976), sudden and unexpected climatic fluctuations are the greatest climatic hazard to agricultural production, and year-to-year variability can be considered that type of fluctuation. It appears that such fluctuations do occur, but the areas and times of occurrence have little predictability. This makes it very difficult to establish and maintain a uniform yearly rate of food production.

Expected Future Changes

Any projections of future climate changes must be considered as speculation. Perhaps the most elaborate recent estimates were projected by the National Defense University (1978a), which conducted a survey of expert opinion on climate change to the year 2000. To some, this is "science by consensus," but it seems worthwhile to examine the opinions of the experts in the field. These experts stated that the causes of global climate change remain in dispute. The experts' answers to the questions indicate a range of opinions. The salient finding was that the likelihood of a catastrophic climate change by the year 2000 is small. More specifically, the responses to the survey suggest only 1 chance in 10 that average global temperature in the next 25 years will increase by more than 0.6°C relative to the early 1970s. Likewise there is only 1 chance in 10 that it will decrease by more than 0.3°C. The most likely event will be a climate that resembles the average of the past 30 years, arising primarily from a balancing of the warming effect of carbon dioxide with the cooling effect of a natural cooling cycle. However, the respondents tended to appreciate a slight global warming rather than a cooling.

The panelists also perceived that any global temperature changes would be amplified at higher latitudes, particularly in the Northern Hemisphere. For example, for the large global cooling scenario, 80 percent of the experts projected a 0.5 to 1.5°C change in both the lower mid-latitudes and higher mid-latitudes, although there was a tendency for values nearer 1.5°C at the higher latitudes. The same percentage of panelists projected a 1.5 to 3.0°C change in the polar latitudes. In the subtropical latitudes, a temperature change of only 0 to 1.0°C was projected by all the experts.

The panelists' responses reflected fairly strong support for the continuation of a 20-22 year drought cycle in the High Plains of the United States. There was less support for cyclic droughts for mid-latitude regions outside the United States. No periodicity was identified relative to frequency of drought in the Sahel region of Africa or the failure of the Asian monsoons.

Collectively, the climatologists expressed considerable uncertainty about possible changes in the amount and variability of precipitation—uncertainty, not only with respect to the magnitude, but in many cases even with respect to the direction of change. This uncertainty was particularly pronounced about possible changes in year-to-year variability. There was some tendency to associate more precipitation and decreased variability of precipitation with global warming and less precipitation and increased variability with global cooling. The effect of less precipitation with global cooling would tend to be detrimental for U.S. agriculture, while the increased precipitation with less variability accompanying global warming would tend to be beneficial.

HUMAN IMPACT ON CLIMATE

Although many of our current activities may affect the climate, it is difficult to project specific effects. Human influence on climate, which already may be appreciable, can only be properly assessed when the natural forces at play are understood (Bryson, 1974). Landsberg (1975) has summarized the causes of climatic alterations, the elements affected, and the "space scale" of alteration (Table 1). No attempt will be made to discuss all the causes; only selected ones will be discussed briefly.

Deforestation and Desertification

Deforestation is occurring in many areas. For example, 60 percent of central Europe has converted from forest to farmland in the last 1,000 years (Sagan, *et al.*, 1979). Major deforestation has taken place in the forested areas of southeast Asia and is now taking place in the Amazon Valley. Deforestation can change the reflectivity of the ground surface, depending on how the land is managed. Changing the reflectivity could affect the climate, but the extent of this is largely speculative. Of greater importance may be the increased susceptibility of the land to wind and water erosion.

Table 1.   Climatic Alterations by Man

Cause	Elements Affected	Space Scale
Deforestation Aforestation	Temperature, evaporation, wind, runoff, erosion	Local Local
Agriculture	Temperature, wind, evaporation, soil blowing	Local to Subcontinent
Drainage	Evaporation, runoff	Local
Irrigation	Evaporation (rainfall?)	Local to Regional
Reservoirs	Evaporation, temperature, (rainfall?)	Local to Regional
Urbanization	Temperature, humidity, runoff, rainfall, air pollution	Local
Industrialization	Air pollution, rainwater composition	Regional
Power Production	Air pollution, thermal pollution, rainwater composition	Regional to Continental
Carbon dioxide production by preceding activities	Temperature	Global
Nitrogen oxides from stratospheric aviation	(Temperature?), ozone reduction, increased ultraviolet radiation	Global
Chloro-fluoro Compounds	Ozone reduction, increased ultra- violet radiation, (temperature?)	Global

SOURCE: Lindsay, 1975

Desertification—the change of steppe regions into desert regions—is another problem. Desertification is the result of an unfavorable climate coupled with poor use of the land. Ensminger (1978) stated "desertification is a greater threat to the world food supply than the population explosion." A large portion of the world's wheat supply is produced in steppe climates. The Food and Climate Forum (Roberts, 1978) stated that each year as many as 14 million acres of productive land turn into desert. This process has been underway and has been accelerating for centuries. For example, the Rajasthan Desert in India has expanded due to overgrazing, and the resulting increase in dust content of the air may further expand the desert. The U.N. Conference on Desertification (1977) stated, "If one degrades a natural ecosystem . . . one simultaneously alters the surface reflectivity. . . . Ecosystem change is inevitably climate change [and] . . . the thermal microclimate may differ by an amount equivalent to many degrees of latitude." This is a people-created problem that appears to be getting worse because of agricultural pressure on the land. Unless current programs and policies are changed, the desertification process will continue.

Atmospheric Particulates

Particulates, fine particles suspended in the atmosphere, change the heat balance of the earth by reflecting and absorbing radiation from the sun and the earth. There are many natural sources of particulates that enter the atmosphere (sea spray, wind-blown dust, volcanoes, the conversion of naturally occurring gases into particles). Indeed, there are more particulates introduced into the atmosphere from natural sources than by mankind (MIT, 1970). However, man does introduce significant quantities of sulfates, nitrates, and hydrocarbons, primarily due to the burning of fossil fuels. In addition, blowing soil due to overgrazing in semiarid regions or high-production farming combined with dry weather periods can introduce large amounts of dust into the atmosphere. Particle levels have been gradually increasing over the years.

Particles in the lower atmosphere can change the earth's reflectivity, cloud reflectivity, and cloud formation. Roberts (1978) points out that Reid Bryson speculates that dust may lead to global cooling (and counterbalance the carbon dioxide warming), but other work suggests it may have some warming effect. At this time it appears that the magnitude of these effects is not predictable and that the direction of the effect (i.e., warming or cooling) has not been determined.

Ozone Content

Changes in the ozone content of the atmosphere have been of considerable concern in recent years. Ozone in the upper atmosphere is extremely important since it partly controls the upper atmosphere's temperature and shields life on earth from harmful ultraviolet radiation, which can cause skin cancer. Changes in the atmosphere's ozone content may induce changes in our weather systems. The ozone concentration in the atmosphere is maintained by a balance between several competing chemical reactions and atmospheric transport processes. Nitrous oxides and fluorocarbons (chlorine is the culprit chemical) are the main pollutants that will reduce the ozone content of the atmosphere. At one time, the nitrous oxides and ozone contents of the atmosphere were in balance, but now that we are introducing nitrous oxides into the atmosphere in amounts equal to that of nature, a new equilibrium balance has to be established. The fluorocarbon input into the atmosphere by the United States has been drastically reduced due to restrictions, particularly on spray cans, but world input still is rising.

Carbon Dioxide Content

We also are increasing the carbon dioxide (CO_2) content of the atmosphere. Roberts (1978) stated that the world-wide atmospheric content of (CO_2) has been increasing at the rate of 0.3 percent per year during this century.

CO_2 is very significant to agriculture, as it is an essential building block in photosynthesis. The process of photosynthesis tends to increase in rate with increased CO_2 content. Agriculture could be affected even more greatly, however, by the meteorological impact that a change in CO_2 content could have. Although CO_2 is a rather poor absorber of short-wave radiation from the sun, it does absorb part of the infrared (heat) radiation from the earth. If this was the only result of increased CO_2 in the atmosphere, the so-called "greenhouse effect" would cause a warming; simple global models suggest that increased CO_2 will produce a global warming of perhaps 2 °C within the next 50 years. This change is greater than any natural fluctuation that has occurred over the past 2,000 years.

An important factor that has seldom been considered in any of the models is a possible change in cloudiness due to increased CO_2 levels. It has been estimated that a 2.4 percent increase in the amount of low cloud over the whole globe could lower the average surface temperature by about 2°C, negating the CO_2 warming. As far as can be seen, there is no simple coupling between temperature and cloudiness. Note also that many factors controlling radiation such as cloudiness and the global albedo (reflectivity), may be undergoing undetected time trends due to natural or perhaps human activities (MIT, 1971). Not all scientists agree that the change will amount to 2°C, but the majority of scientists predict a warming effect of some degree.

Very little research has been done on the possible effects of CO_2 on the water balance. In a recent study, Manabe and Wetherald (1980) used a model that included a simplified interaction between cloud and radiative transfer. Their model showed that a general warming occurred as well as a decreased north-south temperature gradient. An increase in the moisture content of the air also was predicted, but the change did not occur uniformly in different geographical areas. High latitudes showed increased precipitation. A zonal belt around 42° latitude decreased in soil moisture—an event that would severely offset grain production. The model also showed a zone of enhanced wetness along the east coast of the subtropical portion of the continent.

At present, we cannot be sure what the effects of increased CO_2 will be. What we do know is that we will continue to increase the CO_2 content of the atmosphere with the continued consumption of fossil fuels. Such areas as major cities, in which fossil fuels are used in large quantities, will tend to become local heat islands. The local climate may then change because of this heating.

Acid Rain

Industrialization and power production are creating a problem of great concern in some areas of the world—namely, acid rain. Galloway, et al. (1978) stated, "Acid rain is a dominant feature of man-induced change in the chemical climate of the earth. The change is particularly evident in rural and urban areas throughout eastern North America and in many urban areas in the western United States. It has already been recognized as a major environmental problem in northern Europe and Japan."

Acid rain was discovered after the European Chemistry Network was set up in the 1950s. Because the prevailing southerly winds dump western Europe's acidic pollution on Norway and Sweden, the problem became acute early in those countries (West, 1980).

Natural rainfall without man-made pollution has a pH of 5.65 due to natural CO_2 in the atmosphere. Acid rain is due primarily to gaseous man-made pollutants such as sulfur oxides and nitrogen oxides produced primarily, but not exclusively, from the combustion of fossil fuels. These pollutants combine with oxygen in the atmosphere and produce dehydrated sulfuric and nitric acids. A high proportion of nitric acid derivatives in the atmosphere is probably caused by automobile or other mobile sources, while a high proportion of sulfuric acid derivatives is produced by stationary sources such as power plants, smelters, and heavy industry. The latter sources are widely dispersed as a result of actions taken to reduce local pollution. With tall stacks, the material is emitted up to several hundred feet into the atmosphere, carried aloft, and widely dispersed by the movement of the atmosphere. The time for this dispersal, often several days, allows photochemical reactions to take place and convert oxides to acids.

Losses of trout in Norwegian lakes have been noted since the 1920s, but the number of fishless lakes has increased rapidly in the past 15 years, coinciding with the large increase in the combustion of fossil fuels in industrial and urban areas of Europe (Likens, 1976). Likens reports that fishless lakes in Norway now number in the thousands; estimates for Sweden exceed 15,000.

In the United States, the earliest reports on fishless lakes came from the Northeast, with over 100 now reported in the Adirondack area alone (Glass, 1979). Acidity in the Northeast is 60-70 percent due to sulfuric acid and 30-40 percent due to nitric acid. The neutralizing ability of such areas, with a geology that has led to acidic soils and with water bodies low in alkalinity and calcium reserves, is limited, and resistance to further acidification is severely tested by acid rain. Areas of acidic bedrock in upper Michigan, northern Wisconsin, and northern Minnesota are also particularly vulnerable to acid rain.

The extent of the problem can be seen in Figure 4. In 1955-56, the area with pH less than 4.5 was relatively small, and 4.42 was annual average value reported. By 1972-73, that area had expanded greatly, with the lowest value reported as 4.07. Brezonik,

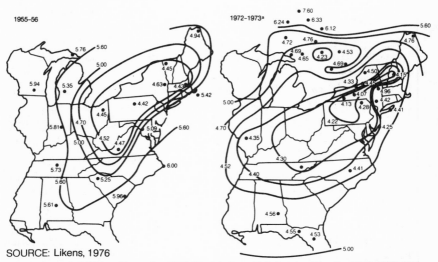

SOURCE: Likens, 1976

a. Data from Oak Ridge, Tenn., for 1973-74; data from Tallahassee, Fla., for 1974-1975; and from Gainesville, Fla., for 1976. Source: Various, including C.V. Cogbill, Thomas Burton, Patrick Brezonik, and Gray Henderson.

Figure 4.   The weighted annual average of pH of precipitation in the eastern United States in 1955-56 and 1972-73

et al. (1980) reported annual average pH values below 4.7 over the northern three-quarters of Florida. Lewis and Grant (1980) reported a significant downward trend in pH near the continental divide in Colorado; increasing acidity is not strictly an eastern U.S. phenomenon. However, the greater concentration of industries in the East makes acid rain a more extensive problem for that area. The ratio of sulfur to nitrogen derivatives indicates that the acidity near and in urban areas in the West is probably attributable to automobiles.

The effects of acid rain are related to both the proximity of pollution sources and the nature and properties of materials affected by acid rain. In the western Prairie region, the soils are calcareous and precipitation would have to become very acidic to significantly offset that area (Glass, 1979).

The effects of acid rain on fish in lakes are well-documented. The effects on stationary structures also are fairly well understood. Acid deposition is helping erode stone monuments and statuary throughout the world. The 2500-year-old Parthenon has shown much more rapid decay in this century than in previous centuries (EPA, 1979).

The effects of acid rain on agriculture, however, are not as well understood. There are two basic ways that acidic precipitation could affect agricultural crops. First, acidic precipitation can impair the surface of the plant. Leaves may become spotted and necrotic. This reaction would be most immediate on crops such as lettuce, chard, and spinach, of which the foliage is the salable portion. Photosynthesis also can be affected by damage to the leaf surface but it is not known how this translates into yield. Second, the plant could be indirectly affected through the soil by changes in the rate at which nutrients are recycled, the speed at which organic matter is decomposed, and the rate at which nutrients are leeched out of the soil. Productivity of both crop plants and forests could be affected. A recent EPA study points out that certain food crops may be benefited by an acid soil—for example, tomatoes and strawberries. In general, however, the effects would be more detrimental than beneficial. A national program for assessing the extent of the problem of acid rain is now underway (Galloway, 1978).

Weather Modification

Intentional weather modification is too complex a subject to be adequately discussed here. There are several difficulties that reduce the potential for cloud seeding to respond to fluctuations in the climate. The space and time variability of precipitation falling over a region is so large that proof of small changes in precipitation is almost impossible. Cloud seeding only augments natural rainfall; it does not stop a drought. The most favorable areas for weather modification seem to be in the mountains, where winter snow pack can be increased. If it were available, this water could be beneficial for agriculture.

Unintentional modification also may be important. In the section on carbon dioxide, a "heat island" effect was mentioned. The precipitation pattern may also be modified around cities (Changnon, 1974) because of inadvertent cloud seeding. Changnon reported average increases of precipitation of 14 percent due to urbanization.

EFFECT OF CLIMATE CHANGE ON YIELD

Since in most cases we cannot predict specific climate changes, we can only estimate what the yield impact might be if a certain event or change occurred. We must recognize that there are difficulties in estimating impacts on crop yields, even if we know what change might occur. In my opinion, one of the major problems is estimating the degree of climate change that would occur at different times of the year with an annual change of a determined amount. In most cases, it is assumed that the change will take place equally throughout the year. We know, as mentioned earlier, that the change will not occur equally by latitudes. One might well argue that change also will not occur equally throughout the year. For example, as indicated previously, Yarger (private communication) found that the cooling trend in Iowa since 1940 was largely a January event. Since corn and soybeans are not grown much at that time of the year, the cooling trend could not substantially affect crop yield.

A cooling trend spread equally throughout the year might be beneficial if it reduced high summer temperatures, or detrimental if it shortened the growing season. It is difficult, however, to evaluate these different effects. There is little or no evidence concerning the seasonal pattern of past changes, so we are speculating when we estimate how any future changes might occur. The wrong assumption could completely invalidate a yield projection, and I cannot overemphasize the importance of this factor.

Alternative Weather Scenarios

This discussion on possible yield changes will be based primarily on material from the Institute of Crop Ecology (1976) and the National Defense University (1978b).*

Four weather scenarios were used in the Institute of Crop Ecology study: (a) the 1933-1936 scenario, a severe climatic stress period, (b) the 1953-1955 scenario, a moderate stress period, (c) the 1961-1963 scenario, a high yielding period, and (d) the 1971-1975 scenario, a period when yields were both above and below the trend line. The results of the study can be shown best by using charts from the report. Figure 5 shows each scenario for Canadian wheat production. The horizontal line is a reference line showing the 1975-1976 level of Canadian consumption (or "dis-

---

*The first report represents a six-month study during which several past weather scenarios were examined. Impacts were studied using actual weather data for selected years but assuming constant 1975 crop area and constant 1973 technology for all years. The second report represents the consensus opinion of a group of experts who considered the impact of selected climate changes on the crops produced in major crop areas of the world.

appearance" in the terminology of the U.S. Department of Agriculture [USDA]). Canadian wheat fluctuated widely in its annual production but in all years was projected above the consumption level. U.S. wheat production (Figure 6) showed less variation than Canadian wheat and was well above the U.S. consumption level

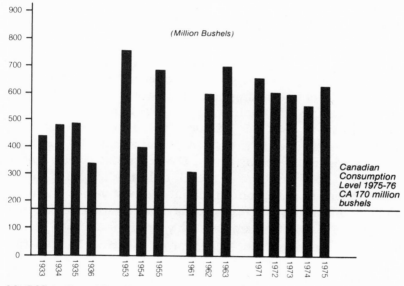

SOURCE: Institute of Crop Ecology, 1976

Figure 5.   Scenario for Production of Canadian Wheat

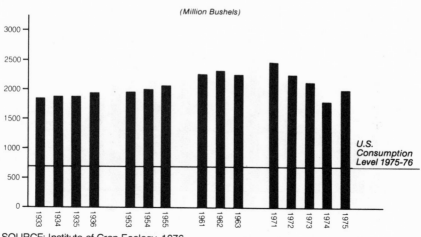

SOURCE: Institute of Crop Ecology, 1976

Figure 6.   Scenario for Production of U.S. Wheat

each year. Corn production (Figure 7) fluctuated widely and was projected below consumption levels in certain years.

Sorghum yields also have fluctuated widely and were projected below the consumption level in one year. Soybean yields were shown to be quite stable for the different scenarios.

Figure 8 shows how corn yields have fluctuated with good weather years and bad weather years. Year to year variability can be large; for example, yields can vary by as much as 20 percent between consecutive years.

If the U.S. corn, wheat, and sorghum yields are combined, year to year variability declines (Figure 9). None of the years produced less than the level of consumption. (This would assume one crop could be substituted for another.) The figure implies that, even in bad drought years, total U.S. production would exceed consumption. However, if added to the U.S. consumption is a total of four billion bushels (B. bu) exported in 1979—corn (2.4 B. bu), wheat (1.325 B. bu), and sorghum (0.275 B. bu)—the total use would exceed combined production for all but the best years. The total production during sequences of good and bad weather years could vary tremendously. For example, each year of the 1961-63 scenario produced more than the best year of the 1933-36 scenario,

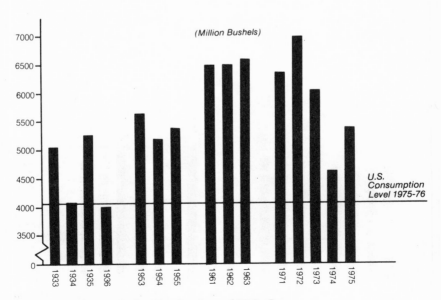

Figure 7.   Scenario for Production of U.S. Corn

SOURCE (FIGURES 7, 8, AND 9): Institute for Crop Ecology, 1976

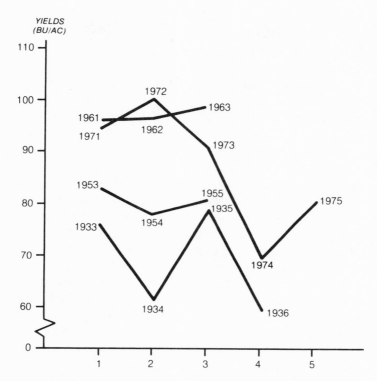

Figure 8. Year to Year Variability of Corn Yields for Different Scenarios

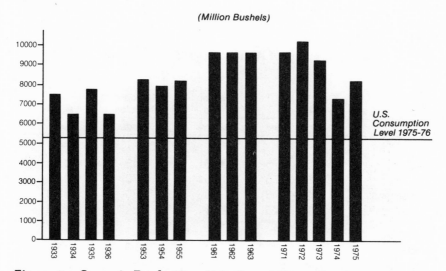

Figure 9. Scenario Production, U.S. Corn, Wheat, Sorghum Combined

obviously creating a large difference in total production over different periods of years. This example demonstrates why surpluses can vary so much in different periods.

Crop Yield Variability in North America

An analysis of the variability of corn yields for all years, starting in the 1930s, was made using a 10-year moving average analysis (Institute of Crop Ecology, 1976). The relative variation (using coefficient of variation—standard deviation/mean) was greater in the 1930s and 1950s than in the 1940s and 1960s. This was not true for the absolute variation. Using the standard deviation as a measure, the absolute variation was greatest in the 1960s.

Comparisons of yield variability among crops show a definite difference in variability among crops (Table 2). Canadian wheat shows the greatest variability. U.S. corn (standard deviation) or sorghum (coefficient of variability) are the next most variable. Soybeans are the least variable, followed closely by U.S. wheat. The wide range of latitudes over which U.S. wheat is grown provides a climatically heterogeneous area. Variability of spring wheat yields is similar to the variability in Canadian wheat yields.

Although the relative variation has decreased, the absolute variability of the major North American grain crops has increased over the past 50 years. However, to quote from the Institute of Crop Ecology report, "Placing an acceptable statistical estimate of the reduction in relative variability over the past three or four

Table 2.   Measurements of Annual Crop-Yield Variability

Crop Species	Period Record	Mean Yield	Standard Deviation	Coefficient of Variability
	Crop years	Bu/acre	Bu/acre	Percent
U.S.				
Corn	1866-1975	35.6	4.46	12.5
Wheat	1866-1975	16.5	1.70	10.3
Sorghum	1929-1975	28.5	3.80	13.3
Soybeans	1924-1975	20.1	1.50	7.5
Canadian				
Barley	1922-1975	27.4	4.50	16.4
Wheat	1922-1975	18.6	4.30	23.1

SOURCE: Institute of Crop Ecology, 1976

decades is not the difficulty. It is, rather, the assessment of the cause, whether it be *good weather* in recent decades or *modern technology* buffering the impacts of bad weather. Within the data sources available, the two are interrelated, therefore inseparable."

Climate Effects on Worldwide Yields

The National Defense University study (1978b) covered fifteen country-crop combinations, as shown in Table 3. These include most of the major grain producing areas of the world.

The agricultural panelists were asked to make point estimates of changes in crop yields due to specified changes from normal conditions of mean crop-season temperatures and precipitation. The reader should particularly note the period for which temperature is considered (only reproduction-period temperature). No consideration is given to possible temperature effects at other times of the year or to any changes introduced into the length of the growing season. No change from the current technology being used in each country was assumed. Taking into account the expressed expertise of each individual panelist for each crop, the reports were summarized and figures prepared showing predicted yield responses. The results for U.S. corn are shown in

Table 3. Country-crop Combinations According to Latitude Zones

Latitude Zone	Crops				
	Corn	Rice	Soybeans	Spring Wheat	Winter Wheat
Northern-higher Middle				Canada U.S. USSR	USSR
Northern-lower Middle	U.S.		U.S.	PRC	U.S.
Southern-lower Middle	Argentina				Argentina Australia
Northern Subtropical		India PRC			India
Southern Subtropical		Brazil			

SOURCE: National Defense University, 1978b

Figure 10. The vertical axis measures, in Celsius degrees, the departure of the temperature (T) of the reproduction period from the long-term average for that period in the crop region. The horizontal axis measures in percent the departure of crop-year precipitation (P) from the long-term average precipitation. The curves connect annual combinations of T and P, which produce the same normalized relative yield for the region. For example, 90 means a yield 90 percent of the long-term average yield for that area. The point(s) of maximum yield are indicated by an (x). The maximum yield that the panelists could indicate was possibly limited by the boundary condition imposed on the upper limit of the precipitation increase. The National Defense University study truncated any precipitation increase at +80 percent.

The contour graph shows that maximum yields for U.S. corn would be expected with temperature about 1°C below normal temperature and rainfall 20 percent above normal. A marked change in yields occurred with precipitation decreasing to below normal and for temperature departing from normal in both directions. This would imply that a cooling period (at reproduction time) would be beneficial but reveals nothing about any possible effect of early season temperatures on the length of the growing season. A warming trend (at reproduction time) would adversely

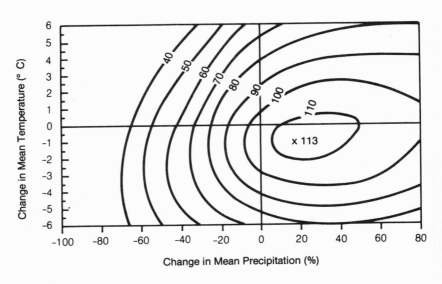

SOURCE: National Defense University, 1978b

Figure 10.   The Yield Contour Graph for U.S. Corn

affect corn, with effects at other times of the season not considered.

Argentine corn also showed increased yield with cooler and wetter weather but with a greater percentage response. Argentine corn would be particularly sensitive to a one standard deviation change in precipitation.

The contour graph for Canada spring wheat is shown in Figure 11. Weather about 3°C warmer and much wetter was considered optimum, with a maximum yield of 144 percent of normal. Cooler temperatures would cause a large drop in yield, as would decreasing precipitation. One standard deviation in precipitation caused significant yield changes. The northern higher-middle latitudes (above 45° latitude) would be benefited by a small warming trend, especially if it included increased precipitation. A cooling trend would be very detrimental. U.S. spring wheat showed a maximum yield with no temperature change but a 60-80 percent increase in precipitation. Decreased precipitation would be very detrimental. A one standard deviation precipitation change caused significant yield fluctuations. USSR spring wheat was similar to that of Canadian wheat, except the maximum yield (145 percent) was obtained with no temperature change but an 80 percent precipitation increase. Both decreased temperature and precipitation

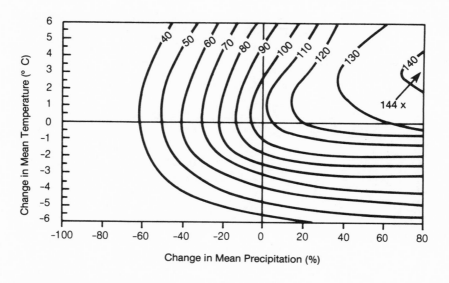

SOURCE: National Defense University, 1978b

Figure 11. The Yield Contour Graph for Canada Spring Wheat

were very detrimental. Argentine wheat showed maximum yields (119 percent) with a decrease of about 2°C in temperature and a 10 percent increase in precipitation and was very responsive to a one standard deviation precipitation change. Australian wheat showed little temperature response but increasing yields with increased precipitation up to 156 percent with an 80 percent in precipitation.

Only the People's Republic of China (PRC) showed an optimum yield at a temperature much different from the present. The experts involved in the study probably knew less about this area than about any other. Whether or not this accounts for their different evaluation of temperature cannot be determined at present. The optimum yield (144 percent) was with an increase in temperature of 3°C and an increase in precipitation of 80 percent. Cooler temperatures were generally detrimental. U.S. winter wheat showed a maximum yield of 126 percent with no change in temperature and a 40 percent increase in precipitation. Decreasing precipitation was very detrimental. USSR winter wheat showed a maximum yield of 131 percent, with no temperature change and an 80 percent increase in precipitation. Decreased precipitation also was very detrimental. For both the U.S. and USSR, decreasing temperature with above normal precipitation was detrimental. The majority of winter wheat areas were most responsive to a one standard deviation precipitation change.

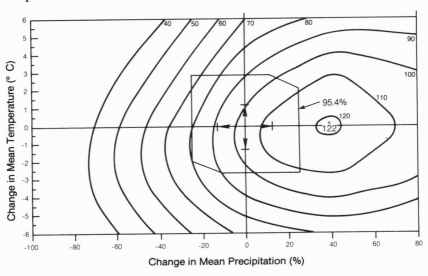

SOURCE: National Defense University, 1978b

Figure 12.   The Yield Contour Graph for U.S. Soybeans

Both Brazilian and U.S. soybeans had similar responses. The maximum yield was with no change in temperature, but with a 20 percent increase in precipitation in Brazil and a 40 percent increase in precipitation in the U.S. (Figure 12). Decreased precipitation decreased yields, and changes in temperature above or below normal, which occurred with increased precipitation, were detrimental.

India and PRC rice showed similar responses with maximum yields occurring with no change in temperature but with a 20 percent increase in precipitation for the PRC and a 40 percent increase for India (very few of the panelists considered themselves experts in this area). Normal temperatures are about optimum. Little temperature change would be expected in subtropical latitudes in either a warming or cooling global trend, so changes in yield should be small. Only with very large temperature changes would yields be reduced. Decreases in precipitation would result in major decreases in yield. Yields were very responsive to a one standard deviation precipitation change.

The panelists also were asked to evaluate the yield effect if a large cooling or large warming occurred in each of the crop-country combinations. Thompson (1979) has summarized these reports as shown in Figure 13. The crop that is affected most is

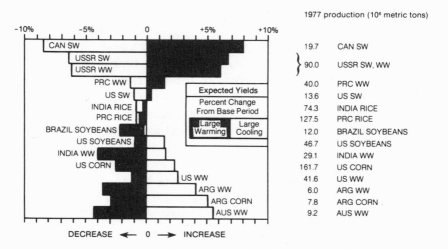

SOURCE: Thompson, 1979 (estimated yield change), and
USDA, 1978 (actual 1977 production)

Figure 13. Estimated Yield Change Assuming a Large Temperature Change in the Climate and Actual 1977 Production Values

spring wheat grown at the higher-middle latitudes. Spring wheat would be favored by a warming trend and adversely affected by a cooling trend in Canada, the Soviet Union, and the United States. Winter wheat in Argentina, Australia, India, and the United States would gain in a large cooling and lose in a large warming. Winter wheat in Russia and China would respond in an opposite manner. A major cooling trend would result in a net decrease of wheat production of about 5.9×10^6 metric tons, or 2.8 percent of annual total production for the countries covered. A major warming trend would result in a wheat increase of about 5.7×10^6 metric tons. Corn is affected less than wheat but, in both Argentina and the United States, the effects are in the same direction. A major cooling would result in a net gain of about 4.6×10^6 metric tons (2.7 percent of total production), while a major warming would result in a net loss of about 4.4×10^6 metric tons. Soybean production would be increased about 0.8×10^6 metric tons with a major cooling and decreased about 0.7×10^6 metric tons with a major warming. Rice was projected to have minor decreases with either a warming or a cooling trend.

Grain Production in the Year 2000

The Foreign Demand and Competition Division of USDA has a computer model for estimating grain production in the world. By making assumptions about technology, an estimate can be made of world grain production by the year 2000 using the previously discussed scenarios. Thompson (1979) has reported on such an analysis; the results are shown in Table 4. It was estimated that world grain production would be about 2,069 million tons in the year 2000 with no change in climate. Present production is about 1,400 million tons. The results of the study indicate there would be little change in total grain production in the world with the various scenarios, but there would be major differences among different countries. The different values shown in the table are probably well within estimation errors involved in the study. I would interpret the table to imply basically no change in total production. The table does indicate, however, that a warming trend would be detrimental to the United States and a cooling trend beneficial. Conversely, the USSR would be affected most by large changes and would benefit greatly by a large warming.

An as yet unpublished report (National Defense University, forthcoming) of Phase II of the National Defense University study showed results similar to those reported by Thompson (1979). Several key points in the summary are listed below:

- The climate changes have the greatest impact in the northern higher-middle latitudes, where global temperature changes are amplified. The Canadian and Soviet wheat crops suffer "large" or "moderate" losses in the cooling scenarios and enjoy similar gains in the warming scenarios. U.S. spring wheat responds in the same directions, but its yield changes are "small."

- The next most sensitive after Canadian and Soviet wheat are the key crops of the Southern lower-middle latitudes, but the directions of their yield responses are contrary to those in the northern higher-middle latitudes.

- Yield changes for key crops of the northern lower-middle latitudes are "small" in all cases. Changes are in the same direction as in the southern zone, except for Chinese winter wheat, which responds like the more northerly wheat crops.

Table 4. Net Changes in Total Grain Production in the Year 2000 with Four Scenarios for Climate Change

	Large Cooling	Moderate Cooling	Moderate Warming	Large Warming
	1000 Metric Tons			
Benefited by Cooling				
U.S.	+10,412	+ 7,807	- 2,002	- 8,695
Others	+15,180	+ 3,019	- 8,135	-13,285
Benefited by Warming				
Canada	- 3,510	- 1,328	+ 189	+ 3,120
USSR	-18,207	-10,391	+8,698	+18,647
PR China	-3,861	-2,843	+ 206	+2,720
World Total Production	2,071,629	2,065,950	2,068,291	2,071,698

SOURCE: Thompson, 1979

- In the subtropical latitudes, most yield changes are "small" and negative. Indian wheat has a pattern similar to U.S. winter wheat.

- Except for Australian wheat, which was estimated to have only a 10 percent yield increase with perceived changes in technology, technological effects are severalfold larger than the magnitudes for the respective climate-induced changes. Yield increases of 50 percent were projected for Argentine corn, with most technology-induced increases in the 20-40 percent range.

Calculations similar to this for precipitation changes have not been made. The impact might well be much greater. At present, however, there are no good estimates of the quantitative change in precipitation to be expected with a climate change, so any results would be purely speculative.

It is difficult to comment on values such as those given in Table 4 unless completely familiar with all the techniques and assumptions used. Several questions obviously could be asked. For example, how does one account for the effect of a climate change on a rich, deep soil, such as in the Corn Belt, versus the poorer soils of the colder regions, such as Canada or the USSR? Soils have been developed over thousands of years. Short-term climate changes have basically no effect on soil properties, thus, production would be controlled by the current production capacity of the soils and the best management that could be applied to those soils. Are the poorer soils capable of responding significantly to much improved weather conditions? Will they respond the same to a new set of climate conditions as they have to the current climate? These would be very difficult answers to obtain, but must be considered in evaluating the effects of climatic change.

CONCLUSION

In his plenary address to the World Food Conference of 1976 (World Food Conference Proceedings, 1977) Clifton Wharton, Jr. said, "The U.S.A. and a few other food surplus nations have the awesome power to decide the fate of the human masses of the world. The U.S.A., for example, has a larger proportion of the world grain markets than the Arab countries have of the world energy market."

About 75 percent of the exportable wheat and coarse grain in the world has come from countries that lie within the temperate climate zone (McQuigg, 1979). It is likely that the temperate zones of the world will continue to be the major source of exportable grain for years to come. What happens to the climates of these zones in the middle latitudes, and particularly the climate of the United States, will be very dominant in determining what happens to the food supply of the future.

Records show that climate has varied in the past, and significant fluctuations have occurred in recent history. There seems to be no reason not to expect fluctuations in the future. The problem is that, with the present state of knowledge about climate change, the climate of the future cannot be predicted. The extent of the uncertainty can be seen from Mitchell (1977), who pointed out four factors related to climate change.

- Future climate will be variable. The only question is whether future variability will be more, less, or the same as the variability of recent years.

- Scenarios used in any study are only examples; what really happens to our climate will probably be different from any of the scenarios.

- What has happened can happen again.

- What will happen in the future is not necessarily limited to what has happened in the past.

It does appear that future agriculture will be operating under climate conditions that cannot now be predicted. The National Defense University study referred to previously (1978b) attempted to evaluate the worldwide effect of temperature changes on crop yield. The net effect was little change in total world production, with some areas showing increases, others showing decreases. The estimates showed that the United States would benefit by a large cooling trend (a decrease of 1°C in the higher-middle latitudes) but would find a larger warming trend (increase of 1.4°C in the higher-middle latitudes) to be detrimental. Crop areas would shift somewhat, but the same crops would be grown.

A climate change, however, may not be the major problem we have to deal with. According to a NRC report (1975), "It is not primarily the advance of a major ice sheet over our farms and citrus that we must fear, devastating as this would be, for such changes take thousands of years to evolve. Rather it is persistent changes (frequent departures above and below normal) of temperature and rainfall in areas committed to agricultural use, changes in frost content of Canadian and Siberian soils, and changes of ocean temperatures in areas of high nutrient production that are of more immediate concern." Or simply put, as stated earlier, it is the wiggles in the climate, not the trends, that are of most concern to us in the immediate future—relatively small areas of the world having a significant fluctuation of their climate, whether due to a real climate change or simply the fluctuation of an unstable climate system. For the present, we must be more concerned about year-to-year variation than long-term changes.

The world's need for food increases every year. In 1976, food for four billion people was required; in the year 2000, we will need food for at least six billion people. Any food crisis in the future may well involve more hungry people than a similar crisis in the past. In our quest for productivity, we have expanded the boundaries of agricultural production to the limit imposed by the average of the climate. We have done our best to optimize crops (varieties and types) to make the maximum use of the resources of soil and climate. In doing this, we have pushed agriculture into more marginal climatic areas where wide fluctuations in production are a fact of life. In spite of modern-day technology, if the weather becomes bad enough, a crop failure will result.

In addition to natural forces, we are adding additional variables to an already complex climate system. We are adding carbon dioxide to the atmosphere by the combustion of fossil fuels. The majority opinion is that this will cause a warming trend in our climate. The ozone content is being affected. Particulate matter from industrial pollution in developed nations and slash and burn agriculture in some developing nations is being added to the atmosphere. The possible effects of this seem "undecided" at present. Large amounts of heat are being added to the atmosphere in selected areas. Overgrazing and forest removal are changing the reflecting properties of the earth's surface. Acid precipitation has become a problem due to atmospheric pollution, and the problem areas are expanding. There seems to me to be little hope of significantly reducing the oxides of nitrogen and sulfur that are causing this problem. With an increasing population and continued high-energy consumption over the world, it appears that we are bringing ourselves closer to the time when the impact of our activities will vie with natural climate forces in determining the course of future climates. It appears that we are adding more uncertainties to an already uncertain system.

In *Living with Climate Change, Phase II*, (J.M. Mitchell, Jr., 1977), it was stated, "There is no question that a repetition of the scenario (1933-36 drought in the Great Plains) would have a severe impact on U.S. agriculture production." But, as the Institute of Ecology-Kettering Foundation study and others have pointed out, some of the changes in technology over the past 40 years would tend to dampen those adverse effects.

I agree with that conclusion in general but believe it requires some qualifications. Technology has allowed production to reach levels considered impossible not too many years ago. The combination of factors such as better crop varieties and increased use of fertilizers has produced a sronger, deeper-rooted plant that is better able to withstand stress conditions. This is the combination that produced the Green Revolution. Under good weather conditions, production is much higher. Under conditions of mild to moderate stress, improved technology has helped to ameliorate stress conditions. Yet, in central Iowa in 1977, we lost a lot of our corn crop to a severe drought. The only thing that would have saved that crop was irrigation. Fortunately, we seldom have a drought of that intensity, but when it occurs, all the technology we have will not save the crop (excluding irrigation). In areas where little or no yield reduction might occur with a mild stress condition, given our present technology, a very severe drought can give us greater yield reductions than we have had in the past simply because we are operating at a higher-base yield level. I would expect to have fewer years with stress reductions than in the past, but in severe years the reduction could be greater. Without irrigation, crop failures are still possible in many areas of the world.

What should we do to provide food for the future? At a meeting I attended several years ago, which considered the Sahelian situation, the question was raised: "Should these areas set limits on animal population that can be sustained over years?" A very good idea in my opinion but an impossible objective. The human drive to improve one's lot is a reasonable objective. But with more people, in a limited agricultural capacity situation, how is it done? Food and Climate Forum (Roberts, 1978) reports the view of a herdsman in the Upper Volta. When asked how he was affected by the drought, he replied he had lost 50 of his 100 cattle. "But," he said, "next time I'll be ready. I'll have 200 to begin with and end up with 100." What he did not realize is that he might well end up with 50, or fewer, because of the greater grazing pressure, and the fact that the land would have suffered even more in the process. Under limiting climates, the land has only limited production potential. Ignorance of proper land management principles and of the consequence of an inappropriate practice are a definite problem in marginal climatic areas. However, these problems are not restricted to such areas. Recently, in the United States, we have seen examples of extreme wind and water erosion due to more land being brought under cultivation or more intensive agricultural practices applied.

The solution is not just a technical agricultural problem. Improved varieties, management practices, etc., can be developed. The solution is also very closely linked to the social-political-economic systems. Do we maintain a food reserve for bad weather years? Who pays for it? How do we maintain production in the United States? At what costs? These are questions needing answers.

As a climatologist, I say we must incorporate climatic data into long-term planning in all areas of the world and use what information we have for planning; otherwise, a run of bad-weather years will produce food problems even more serious than we have had in the past. The question that really is bothersome is this: What would be the result if a combination of events happens such that a major drought occurs in a large area of the world, at a time when no surplus grain is available in other parts of the world? The thought is unpleasant; the results of such an event would be much more unpleasant. Who would make the triage decisions?

I hope that a statement made by the United Nations a few years ago represents more likely what will happen. The statement said, "For the next decade or so the probability is good that [world] food production, *in total* will keep a half step ahead of population, but, there will be *times* and *places* of critical shortage." The year-to-year weather largely will determine these times and places. In the long run, there may be serious problems in meeting the world food demand. A deteriorating climate for agriculture will further aggravate that problem. But I think it is largely guess work when we attempt to predict the climate in the future and its impact on agriculture.

REFERENCES

Brezonik, P.L.; Edgerton, E.S.; and Hendry, C.D. "Acid Precipitation and Sulfate Deposition in Florida." *Science* 208(1980):1027-9.

Bryson, R.A. "A Perspective on Climate Change." *Science* 184(1974):753-60.

Bryson, R.A. "Ancient Clemes on the Great Plains." *Natural History* 89(1980):64-73.

Budyko, M.F., and Vasicheva. *Inadvertent Climate Modification.* MIT Press Publication 201. Cambridge, MA.: MIT Press, 1971.

Ceres. U.N. Food and Agriculture Organization. New York: FAO, 1977.

Changnon, S.A. "A Review of Inadvertent Nesoscale Weather and Climate Modification and Assessment of Research Needs." Paper presented at the Fourth Conference on Weather Modification of the American Meteorology Society, 1974, Boston, MA. Mimeographed.

Cogbill, C.V., and Likens, G.E. "Acid Precipitation in the Northeastern United States." *Water Resources Research* 10(1974):1133-37.

Davis, N.E. "The Variability of the Onset of Spring in Britain." *Quarterly Journal of the Royal Meterology Society* 98(1972):763-77.

Decker, W.L. "Climate Fluctuations and Agricultural Production." In *Climate Technology Proceedings.* Columbia, MO.: University of Missouri, 1977.

Ensminger, D. Desertification Control Bulletin 1-1. New York: United Nations Environment Programme, 1978.

Environmental Protection Agency. "Research Summary, Acid Rain." EPA-600/8-79-028. Washington, D.C.: EPA, 1979.

Federal Council for Science and Technology. "Report of the Ad Hoc Panel on the Present Interglacial Interdepartmental Committee for Atmospheric Sciences." Science and Technology Office, ICAS 136-F475. Washington, DC.: National Science Foundation, 1974.

Galloway, J.N.; Cowling, E.B.; Gorham, E.; and McFee, W.W. "A National Program for Assessing the Problem of Atmospheric Deposition (Acid Rain)." A Report to the Council on Environmental Quality, National Atmospheric Deposition Program, NC-141. Washington, D.C.: CEQ, 1978.

Glass, N.R.; Glass, G.E.; and Rennie, P.J. "Effects of Acid Precipitation." *Environmental Science and Technology* (1979):1350-55.

Hancock, D.J., and Yarger, D.N. "Cross-spectral Analysis of Sunspots and Monthly Mean Temperature and Precipitation for the Contiguous United States." *Journal Atmospheric Science* 36(1979):746-53.

Institute of Ecology-Charles F. Kettering Foundation. "Impact of Climatic Fluctuation on Major North American Food Crops." Dayton, OH.: Charles F. Kettering Foundation, 1976.

Ladurie, E. LeRoy. *Times of Feast, Times of Famine. A History of Climate Since the Year 1000.* Garden City, NY.: Doubleday and Co. Inc., 1974.

Lamb, H.H., and Woodroffe, A. "Atmospheric Circulation During the Last Ice Age." *Quaternary Research* 1(1970)29-58.

Landsberg, H.E. "The Definition and Determination of Climatic Changes, Fluctuations and Outlooks." In *Atmospheric Quality and Climatic Changes, Papers of the Second Carolina Geographical Symposium,* edited by R.J. Kopec. Studies in Geography No. 9. Chapel Hill, NC.: University of North Carolina, 1975.

Lewis, W.M., and Grant, M.C. "Acid Precipitation in the U.S. *Science* 207(1980): 176-7.

Likens, G.E. "Acid Precipitation." *Chemical and Engineering News* 54(1976):29-44.

Manabe, S., and Wetherald, R.T. "On the Distribution of Climate Change Resulting from an Increase in CO_2 Content of the Atmosphere." *Journal Atmospheric Science* 37(1980):99-118.

Massachusetts Institute of Technology. *Man's Impact on the Global Environment.* MIT Publication 162. Cambridge, MA.: MIT, 1970.

Massachusetts Institute of Technology. *Inadvertent Climate Modification. Report of the Study of Man's Impact on Climate (SMIC).* Cambridge, MA.: The MIT Press, 1971.

McQuigg, J.D. "Climatic Variability and Agriculture in the Temperate Regions." Paper presented at the World Climate Conference, Geneva, Switzerland, 1979. Mimeographed.

Mitchell, J.M., Jr. "Record of the Past, Lessons for the Future." In *Living with Climatic Change, Phase II,* pp. 15-26. McLean, VA.: The MITRE Corp., METREX Division, 1977.

Mitchell, J.M., Jr; Stockton, C.W.; and Meko, D.M. "Evidence of a 22-year Rhythm of Drought in the Western U.S. Related to the Hale Solar Cycle Since the 17th Century." In *Solar Terrestrial Influences on Weather and Climate,* edited by B.M. McCormoc and T.A. Seliga, pp. 125-44. Boston, MA.: D. Reidel Publishing Co., 1979.

Moran, J.M.; Dugas, W.A.; and Olson, R. "Agricultural Implications of Climatic Change." *Journal of Soil and Water Conservation* 32(1977): 80-3.

National Defense University. *Climate Change to the Year 2000.* Washington, D.C.: Fort Lesley J. McNair, 1978[a].

National Defense University. *Crop Yields and Climate Change: The Year 2000.* Washington, D.C.: Fort Lesley J. McNair, 1978[b].

National Defense University. *Crop Yields and Climate Change to the Year 2000.* Washington, D.C.: Fort Lesley J. McNair, forthcoming.

National Oceanic and Atmospheric Administration (NOAA). "The Influence of Weather and Climate on United States Grain Yields: Bumper Crops or Droughts." Committee Report. Washington, D.C.: U.S. Department of Interior, 1973.

National Research Council. "Understanding Climate Change." A Program for Action, Global Atmospheric Research Program Committee. Washington, D.C.: National Academy of Sciences, 1975.

National Research Council. "Climate and Weather Fluctuations and Renewable Resources Committee, Board on Agriculture and Renewable Resources. Washington, D.C.: National Academy of Sciences, 1976.

Roberts, W.O. "Climate in Perspective." In *Food and Climate Review, 1978,* pp. 3-8. The Food and Climate Forum. Boulder, CO.: Aspen Institute for Humanistic Studies, 1978.

Rockefeller Foundation. "Climate Change, Food Production, and Interstate Conflict." A Bellagio Conference, 1975. Mimeographed.

Sagan, C.; Toon, O.B.; and Pollach, J.B. "Anthropogenic Albedo Changes and the Earth's Climate." *Science* 206(1979):1363-68.

Science Council of Canada. *"Living with Climatic Change."* Proceedings, Toronto Conference Workshop, November 19-22, 1975. Toronto: Science Council of Canada, 1976.

Starr and Oort. *Living with Climatic Change, Phase II.* McLean, VA.: The MITRE Corp., METREX Division, 1973.

Thompson, L.M. "Weather and Technology in the Production of Corn in the U.S. Corn Belt." *Agronomy Journal* 61(1969):453-6.

Thompson, L.M. "Cyclical Weather Patterns in the Middle Latitudes." In *Soil and Water Conservation* 28(1973):87-9.

Thompson, L.M. "Climate Change and World Grain Production." Prepared for the Chicago Council on Foreign Relations. Chicago, IL.: Chicago Council on Foreign Relations, 1979.

United Nations. *Background Document: Climate and Desertification.* Conference on Desertification, 1977.

U.S. Department of Agriculture. *Agriciultural Statistics 1978.* Washington, D.C.: USDA, 1978.

Van Loon, H., and Williams, J. "The Connection Between Trends and Mean Temperature and the Circulation at the Surface: Part I. Winter." Boulder, CO.: National Center for Atmospheric Research, 1976.

Weakly, H.E. "History of Drought in Nebraska." In *Journal of Soil and Water Conservation* 17(1962):271-5.

Weinbeck, R.S. "Periodic Variability from Long-term Weather Records." Ph.D. Dissertation. Iowa State University, Ames, IA., 1980.

West, Susan. "Acid from Heaven." *Science News* 117(1980):76-8.

World Food Conference of 1976. *The World Food Conference of 1976 Proceedings.* Ames, IA.: Iowa State University Press, 1977.

The Authors

Sandra S. Batie

Sandra S. Batie is an agricultural economist and a Senior Associate with The Conservation Foundation. She is currently directing a research project on soil erosion and conservation in the United States. Dr. Batie's interests are in land and natural resource economics, and she has published extensively in these areas. She received her Ph.D. in Agricultural Economics from Oregon State University. Before coming to The Conservation Foundation in January 1980, Dr. Batie was Associate Professor of Agricultural Economics at Virginia Polytechnic Institute and State University, Blacksburg, Virginia. She is currently on leave from the faculty of VPI and SU.

Otto C. Doering

Otto C. Doering is Associate Professor of Agricultural Economics at Purdue University. He is a public policy specialist and coordinator of energy programs for the School of Agriculture at Purdue. He has written extensively on energy policy for both professional journals and a wider public. Dr. Doering's present activities include a position on the Department of Energy's National Energy Extension Service Advisory Board; he also has served as visiting Policy Economist for the Economic Research Service, USDA. Dr. Doering has a Ph.D. from Cornell University in Agricultural Economics and an M.S. degree from the London School of Economics in Economics and Public Administration. Before joining the faculty at Purdue, he was on the staff at the College of Agriculture in Malaya, and at the Ford Foundation.

Kenneth D. Frederick

Kenneth D. Frederick is a Senior Fellow and Director of the Division of Renewable Resources at Resources for the Future. His research has focused on agricultural development, and he has written numerous articles on water management in agriculture. Dr. Frederick received a Ph.D. in Economics from the Massachusetts Institute of Technology. He was formerly Assistant Professor of Economics at California Institute of Technology, and has been a consultant with the National Security Council and the Agency for International Development.

Jack R. Harlan

Jack R. Harlan is Professor of Plant Genetics in the Crop Evaluation Laboratory of the Agronomy Department, University of Illinois at Urbana. He has written numerous professional papers and several books in the field of plant genetics. His recent work has centered on crop evolution, especially among cereals. Dr. Harlan is a member of a "Panel of Experts on Plant Exploration and Introduction" for the Food and Agriculture Organisation of the United Nations, a Fellow of the American Association for the Advancement of Science and of the American Society of Agronomy, and a member of the National Academy of Sciences. Dr. Harlan received a Ph.D. from the University of California at Berkeley in Genetics. He was formerly with the Agricultural Research Service, USDA, and a Professor of Genetics at Oklahoma State University. He has received the Crop Science Award and the International Agronomy Award of the American Society of Agronomy, and other scholarly awards.

Robert G. Healy

Robert G. Healy is an economist and a Senior Associate with The Conservation Foundation. He is currently director of the Foundation's Rural Land Market Project and co-authoring a book on the rural land market in the United States. The author of numerous articles in professional journals, he is co-author of *The Lands Nobody Wanted: Policy for National Forests in the Eastern United States,* author of *Land Use and the States,* and editor and contributing author of *Protecting the Golden Shore: Lessons from the California Coastal Commissions.* Dr. Healy received a Ph.D. from the University of California at Los Angeles. Formerly on the research staffs at the Urban Institute and Resources for the Future, he also taught city planning at Harvard University.

Philip M. Raup

Philip M. Raup has been a Professor in the Department of Agricultural and Applied Economics at the University of Minnesota since 1953. He teaches courses in land and agricultural economics and his research has encompassed many issues in agricultural economics. He has served as a policy consultant and a member of numerous professional associations. Dr. Raup received his Ph.D. in Agricultural Economics from the University of Wisconsin. Before joining the faculty at the University of Minnesota, he was a research fellow with the Brookings Institution, Assistant Professor of Agricultural Economics at the University of Wisconsin, and Guest Professor at the Institute of Foreign Agriculture, Technical University of Berlin.

Vernon W. Ruttan

Vernon W. Ruttan is a Professor in the Department of Agricultural and Applied Economics at the University of Minnesota. His contributions to the field of agricultural economics have included the induced innovation model of agricultural development, and he has directed extensive research on the contribution of research to productivity growth in agriculture. He has written and edited numerous books and journal articles in agricultural economics and has been a fellow and president of the American Agricultural Economics Association. Dr. Ruttan received a Ph.D. from the University of Chicago. At the University of Minnesota, he served as Chairman of his department, and as Director of the Economic Development Center.

Robert H. Shaw

Robert H. Shaw is Distinguished Professor of Agricultural Climatology in the Department of Agronomy at Iowa State University. Since 1948, he has been in charge of the research and teaching program in agricultural climatology. He has published over 150 articles on many aspects of his field. He is a member of the American Association for the Advancement of Science and the American Society of Agronomy. Dr. Shaw received a Ph.D. in Agricultural Climatology and an M.S. in Plant Physiology from Iowa State University. His activities as a climatology expert have taken him to many parts of the world and have included working with the World Meteorological Organization and the National Research Council, as well as consulting for the Presidential Advisory Committee on Weather Control.

Frederick N. Swader

Frederick N. Swader is Associate Professor of Soils Science Extension at Cornell University. His special focus has been on soil water management, and he has written extensively for both professionals and farm operators on this subject. He spent 1979 as a consultant to the Water Planning Division of the Environmental Protection Agency, where he advised EPA on technical aspects and policy implications of water quality improvement strategies. Dr. Swader holds a Ph.D. in Soil Science from Cornell University, where he has been on the staff since 1967.